QUALITATIVE
MEDIA
ANALYSIS

DAVID L. ALTHEIDE
Arizona State University

Qualitative Research Methods
Volume 38

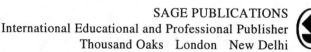

SAGE PUBLICATIONS
International Educational and Professional Publisher
Thousand Oaks   London   New Delhi

For information address:

SAGE Publications, Inc.
2455 Teller Road
Thousand Oaks, California 91320
E-mail: order@sagepub.com

SAGE Publications Ltd.
6 Bonhill Street
London EC2A 4PU
United Kingdom

SAGE Publications India Pvt. Ltd.
M-32 Market
Greater Kailash I
New Delhi 110 048 India

Printed in the United States of America

Library of Congress Cataloging-in-Publication Data

Altheide, David L.
    Qualitative media analysis / author, David L. Altheide.
        p.   cm.—Qualitative research methods; v. 38)
    Includes bibliographical references.
    ISBN 0-7619-0198-1 (cloth: acid-free paper).—ISBN
0-7619-0199-X (pbk.: acid-free paper)
        1. Mass media—Research—Methodology.   I. Title.   II. Series.
P91.3.A48   1996
302.23′072—dc20   95-41796

This book is printed on acid-free paper.

98 99 10 9 8 7 6 5 4 3 2

Sage Production Editor: Gillian Dickens
Sage Typesetter: Andrea D. Swanson

CONTENTS

For Tasha and Tod, who inspire me to teach.

SERIES EDITORS' INTRODUCTION

This is the age of media. Media effects are seen in politics, literary theory, and cultural studies. Although social scientists have been keenly interested in media effects for almost 50 years, they still rely largely on content analyses, surveys, and questionnaires as methods of investigation. Yet modern information technologies are both a topic and a resource for study. Moreover, such technologies are pervasive, cheap, efficient, speedy, and, critically, central sources of modern experience. The understanding of such technologies requires new analytic approaches.

David Altheide presents in Volume 38 of this Sage series a model of meaning within which media documents can be defined and analyzed. *Qualitative Media Analysis* provides a useful logic for organizing media studies and a detailed look at how two exemplary media studies were conducted. Like several other authors in this series—Peter Manning in *Semiotics and Fieldwork* (Volume 7), Paul Atkinson in *Understanding Ethnographic Texts* (Volume 25), and Martha Feldman in *Strategies for Interpreting Qualitative Data* (Volume 33)—David Altheide considers the coding and drawing of theoretically informed inferences from qualitative materials. From practical issues such as how to make videotapes and organize research teams, to analytical issues such as construct development and representation, this monograph provides a guide for the study of just how media documents both reflect and shape social life.

—Peter K. Manning
John Van Maanen
Marc L. Miller

PREFACE

My interest in documents arose from my work with the mass media and particularly television news. My fieldwork and more than two decades of writing about the electronic media convinced me that the visuals and other more subtle images and messages were the most compelling and powerful aspects of their production, as well as what viewers interpreted and took with them. I rejected the assumption that visuals could not be systematically analyzed and set out to contribute a way to study them. But it was not until I began to teach how to conduct media research that I was forced by my students to try to clarify the orientation, perspective, guidelines, and techniques for conducting analysis of TV visuals, as well as the search for meanings and definitions in documents. Ironically, I was often forced to "reconstruct" for students how I did some of the work they were reading for my courses, and, of course, each reconstruction got a bit more clear, more logical, and certainly more "systematic" than the original research! In this sense, I had to learn from the work after it was accomplished, so that I could tell others how to do it, and then I would apply what I said I did to the next project, each time getting rid of a few sharp edges while hopefully sharpening some ideas. The result of this "backing into my next project" is what you see.

This work, as with most of my other efforts, has a very clear bias. It rests upon a "media-centric" view of the world, or the notion that experience is mediated on its way to our meaningful interpretations. And the mass media are a big part of this, but so too are the array of communication logics and formats through which we encounter experience and derive meaning from it. Thus, it is from my media work that I became more aware of the role of communication in everyday affairs. The linkage was the *document.*

A lot of people need to be thanked for help on this emerging project. My colleagues and friends like Bob Snow, John Johnson, and Bud Pfuhl, Jr. provided many ideas, patience, and helped formulate the rationale from which the following pages were cut. The late Carl Couch was continually encouraging in my efforts to make sense of the role of electronic media and formats, yet he was also critical, forcing finer distinctions and analytic approaches. There were also many students who directly and indirectly contributed to this work, several of whom are mentioned in the text, especially Dion Dennis, Robert Todd, Aogan Mulcahy, Robin Rau, and Jennifer Ferguson. Deb Henderson actually read the manuscript when I handed it to her and very irreverently (but helpfully!) made critical comments throughout—suggestions I have tried to incorporate. I would like to thank Janell Nagle and the College of Public Programs Publication Assistance Center for terrific work on the graphics. The three series editors, Pete Manning, John Van Maanen, and Marc Miller, also made challenging suggestions. But it was John Van Maanen who, nearly a decade ago, ran me through the Tempe streets while urging me to do something with qualitative data analysis. My promise about sending him a prospectus was not really broken—it just took a long time. But what the heck, maybe the ideas and approach became more familiar along the way. I hope so, but I thank John just the same.

—David L. Altheide

QUALITATIVE MEDIA ANALYSIS

DAVID L. ALTHEIDE
Arizona State University

1. PLUGGED IN RESEARCH

This book is for students and professionals who seek an integrated approach to the actual conduct of document analysis. Very little of the published work about document analysis is by researchers who actually do document analysis! In the chapters that follow, I draw on 25 years of research experience and teaching research methods to undergraduate and graduate students. Their routine questions and problems regarding conceptualization and data collection guidelines have guided this effort. Indeed, in a few instances, I will use examples from student projects to illustrate points. My main goal is to convey a sense of the "total" research project involving documents so that students can plan, organize, conduct, and analyze their own research projects. Although parts of the package can be used separately, there is a unifying conceptual approach that joins the parts. A related concern is to convey a sense of how this approach to research can be combined with other methods and materials to provide a more complete look at a research topic. This becomes particularly challenging as a host of new electronic information bases are becoming available in libraries as well as home computers. A number of references are included in the Appendix.

This book is intended to fill a gap in research methods between traditional "content analysis" or systematic techniques for the objective study of characteristics of messages (Holsti, 1969) and qualitative methods such as participant observation and focused interviewing. There are numerous discussions of different kinds of content analysis, both quantitative and qualitative, as well as more recent publications about computer-assisted content analysis (Kelle, 1995; Weitzman & Miles, 1995). Several others are listed in the reference section.

My approach is to blend the traditional notion of *objective content analysis* with *participant observation* to form *ethnographic content analysis,* or how a researcher interacts with documentary materials so that specific statements can be placed in the proper context for analysis (Altheide, 1987). The next chapter discusses this approach in some detail, but the immediate concern is to suggest why an expanded view of document analysis is needed.

Documents are studied to understand culture—or the process and the array of objects, symbols, and meanings that make up social reality shared by members of a society. For our purposes, a large part of culture consists of documents. A *document* can be defined as any symbolic representation that can be recorded or retrieved for analysis. *Document analysis* refers to an integrated and conceptually informed method, procedure, and technique for locating, identifying, retrieving, and analyzing documents for their relevance, significance, and meaning. Broadly conceived, all research materials are potentially documents within the researcher's framework. All documents that are selected can be used as data, but not all data are documents. The meaning and significance of all documents is informed by the research perspective and act. Most human documents are reflexive of the process that has produced them. A researcher prefers to be aware of this process to understand the meaning and significance of the document. It is the *researcher's interest and relevance* plus the retrievable characteristic that produce a research document. If something is relevant but not retrievable, it does not qualify as a document, even though it may be helpful for the overall project. The document has an existence independent of the researcher, although its meaning and significance for the research act will depend on the researcher's focus—that is, the document will not be transformed into "data" without the researcher's eye and question. Moreover, researchers' findings and interpretations of the document reflect a perspective, orientation, and approach.

Culture is difficult to study because its most significant features are subtle, taken for granted, and enacted in everyday life routines. Our capacity to study many aspects of culture is closely related to theoretical ideas about what is important as well as technological capacity to capture what we would prefer to examine. Not surprisingly, these are related, because technological innovations expand some of the conceptual questions one can ask. One needs to have the vision to "see" possibilities—these are theoretical insights—but also needs the capacity to pursue this new vision.

This is also true of document analysis. Our current capacity for exciting document analysis surpasses our conceptual awareness of what to do, how to do it, and how to interpret what is found. This is part of an *ecology of communication* of the structure, organization, and accessibility of various forums, media, and channels of information (Altheide, 1995). This means, quite simply, that the increased technological capacity to record and retrieve information has expanded the range of potential documents. For example, the availability of video cameras since the 1970s now makes routine interaction documentable in a way that was not readily available and affordable even with 8-mm and 16-mm film. Whether they be completed surveys, interview schedules, official records and statistics (e.g., the FBI Uniform Crime Report), a newspaper or television (TV) news transcript, photographs, or field notes, these are all documents that the researcher intends to reflect some event or activity within a sphere of interest.

There are three classes of documents relevant to researchers. First are *primary documents,* which are the objects of study. Examples include newspapers, magazines, TV newscasts, diaries, or archaeological artifacts. Most of this book is about this class of documents. Next are *secondary documents,* which are records about primary documents and other objects of research. This includes field notes, published reports about primary documents, and other accounts that are at least one step removed from the initial data source by a researcher or some other "filter." Diaries may also be included in this category. The last category of documents is a catchall, and this I refer to as *auxiliary documents,* which can supplement a research project or some other practical undertaking but are neither the main focus of investigation nor the primary source of data for understanding the topic. These materials are often discovered by the researcher as being relevant to understanding a particular aspect of a study. For example, the contents of a garbage can may be useful to help clarify what a news editor finds as not newsworthy, or in the case of a long-term project on consumer behavior called *garbology,* refuse can clarify actual consumer and eating behavior— for example, "how much alcohol is actually purchased." If one is only studying garbage, of course, then garbage is the primary document. Footpaths worn in grass, dog-eared pages in books, and other unobtrusive indicators can become documents under a researcher's purveyance. Wearing apparel and photographs of how research subjects dress may be useful documents when combined with other data about social class, fashion, and hobbies. This book will not focus on auxiliary documents except to note their importance for most research projects and to encourage researchers to look for such documents to clarify and illustrate data derived from other research materials.

The relationship of these documents in a research project can become a bit complex. On the one hand, an ethnographer studying a TV newsroom obtains data from what is seen and heard working with journalists. There may be written documents that the investigator collects, such as a reporter's notebook or copies of the station's editorial about a mayoral candidate. The latter would be treated as auxiliary documents to the researcher's field notes chronicling observations and interviews, which are secondary documents. On the other hand—and consistent with most of the material in this book—a researcher may investigate newspaper or TV news transcripts and visuals. These are the primary documents from which the researcher may collect data using a protocol or a secondary document, similar to an account of an account.

These brief remarks about a typology of documents should not be regarded as cut-and-dry in that it is ultimately the researcher's perspective and focus that makes particular documents relevant. For example, if one wished to study the interview notes or diaries produced by other researchers to understand their language and assumptions, then those notes would be the primary documents for the investigation. Regardless, a researcher would prefer to have access to as many relevant documents for a study as possible. Moreover, part of the research quest is to actually "discover" new documents that may be treated as auxiliary or supplemental to one's main focus.

Problems With Studying Documents

Research is a social process. The social and cultural environments in which one operates as an investigator contribute to how one views research problems, data sources, and methodological approaches. In addition to political and funding "guidelines" for studying certain kinds of advertising and mass media documents, the way one studies documents has been influenced by context, particularly retrievability and access. For example, many approaches to the study of documents were constrained by more limited documents, information technology, mass media, and popular culture, as well as a positivist model of science (or objective analysis) independent of the investigator's method and theory. The latter point has been exhaustively analyzed and will not be repeated here, except to state simply that most approaches to content analysis were grounded in a tradition that equated "true knowledge" with numbers and measurement and, therefore, collecting quantitative data. Assumptions were made about the media's impact on audience members, who essentially were regarded as being very passive and subject to influence by the bulletlike impact of messages. It was assumed that simply studying the frequency and pattern of bullet-messages would tell us what was happening to audience members. Of course, we now know that audience members are very "active" and interpret messages in many different ways, so the impact of a message cannot be understood without examining other relevant social factors.

Documents subject to analysis were also quite limited, consisting mainly of printed materials (e.g., newspapers, magazines, letters, and diaries). Information technology was significant for document analysis because of the problem of access and retrievability. This was important because the documents were limited by their own physical availability and existence: Only if, for example, newspaper articles could be physically stored somewhere could they be retrieved. Archives of news materials, historical documents, and records became the primary source of data. With the advent of photography, including microfilm, microfiche, and the like, a document's life and access were extended through reproduction and making copies available to libraries.

Most of the reproduction technology, rationale, and procedures were owned and controlled by bureaucracies and formal organizations. This was very important for the thrust, cost, and actual conduct of document research. For example, if a researcher in the 1940s wanted to compare the *New York Times*'s newspaper coverage of homicides of men and women over a 10-year period, he or she would have to gain access to newspaper archives at the *Times*'s offices or meticulously go through daily news reports just to find the news stories! This would take several months and be very expensive. Moreover, just gaining access to the materials could be very difficult, requiring permission from various individuals in the newspaper organization. It would be very unlikely for a graduate, and a virtual impossibility for an undergraduate student, to have this access. If this same project were carried out in the 1960s, a researcher could use microfiche (photocopies) of the *New York Times* in his or her own university or public library, but the project would still take several weeks just to find the homicide articles by going through each issue (if the researcher did not have access to a "newspaper morgue"). Students, however, would be more likely to have access to these materials.

Possibilities for document analysis have expanded geometrically during the past 20 years. What used to be major undertakings to conduct research can now be done by undergraduate students with a little training in a matter of several weeks! Massive changes in information technology and culture were reflected in new documents and popular culture, particularly electronic media. No culture in history has been more recorded or replayed—but not yet analyzed!—than that of the United States (and increasing portions of Western Europe) during the past two decades. The method, approach, and relevance of document analysis changed with movies, radio, audio recordings (records), television (and later, cable TV), videos, video cameras, video recordings, compact discs, computers, computer games, and the range of message delivery systems relevant to the Internet or "information highway." More and more experience, business, news, and fun were being processed as "media" products. Popular culture became a significant part of everyday life. But most important for researchers is that these devices and products became more easily accessible, retrievable, and affordable. For example, using LEXIS/NEXIS (one of the information bases to be discussed in a later chapter), a typical undergraduate student could obtain the data for the comparative crime study noted above in a few days!

With the integration of information technology throughout popular culture, document analysis covers a wider range of topics than previously. When the officials and heads of organizations controlled access to relevant documents, research was severely limited, even to the point of getting approval for a research topic! Relevance was also limited by documents that were retrievable. Archival records severely limited what could be studied, and with the exceptions of some studies using personal letters and diaries, most research using documents available at the time would have benefited substantially from other sources of data that were not so readily available.

Culture is more available to document analysis today partly because the electronic and information technology revolution that is the source of such research is also the single greatest contributor to cultural change. Because of cable TV, syndicated reruns, and videotape, it is now possible to study the same television programs today that had a major impact on how American families thought about themselves (e.g., Ozzie and Harriet) in the 1950s. Indeed, it is partly because such materials are on "record" and have been studied as cultural documents that we have a better understanding of the impact of popular culture on "nostalgia," "cultural myths," and our sense of the future (Combs, 1984). The upshot is that some familiarity with document analysis is useful not only for research questions but also for a kind of cultural self-awareness. Because most of the relatively new information technologies are based on visual and aural (in addition to literal or narrative) materials, I present several illustrations of conceptual and data-gathering approaches appropriate for these documents.

Information technology has opened up a potentially enormous source of new documents for investigation of culture, but this entails adjusting views of documents and their significance as well. A related concern is how changes in information technology, such as computer applications, can direct how we approach documents rather than complementing an approach that may be more theoretically informed. For example, there is a growing supply of computer programs for qualitative document analysis. This has surfaced as more qualitative-oriented researchers provide a potential market but also because working with text materials is very demanding, cumbersome, and seldom as precise and systematic as when one works with quantitative data. I note in a later chapter that such reliance on software that is mainly oriented to helping one code data is not always fruitful for cogent and creative analysis. For that reason, there is very little material in this book about software applications.

An Approach to Document Analysis

The materials in this book are oriented to the analysis of multiple documents rather than one in detail. Although single document analysis is not necessarily incompatible with multiple documents, historically there has been a much different emphasis. The approach to examining single documents (e.g., a movie, a novel, or a McDonald's restaurant) is called *semiotics,* or the study of signs. There are a number of detailed accounts of the use of semiotics in social science research (Berger, 1982; Manning, 1992; Manning & Cullum-Swan, 1994). Such studies focus on depth, and unlike the kind of document analysis I deal with, the critical emphasis is on trying to unravel the author's assumptions, motives, and intended consequences as revealed by analysis of the document. Manning and Cullum-Swan (1994) capture the orientation of a semiotic task, especially if guided by an interactionist perspective, in the following:

> Reading entails an audience. . . . Reading is done by scholars, critics, other writers, reviewers, historians and related intellectuals. The serious critic intends to reconstruct the process of writing, reading, and reflection and to ruminate upon, according to the conventional canons of taste and the genre,

the quality of the writing. The critic's task is to place the writing, the text, and its readings into alternative contexts or fields, or to recode the text. Adequate criticism should enable others to "penetrate" the author's intent and the tenor of the times within which the text existed, to strip away lies and stylistic obfuscations, and to discover therefore the deeper or "real" meaning of a written product. (p. 468)

As with most research approaches, any attempt to look backwards from a text toward the author's motivation is rich with problems, especially if the author lacks an awareness of and familiarity with the historical, cultural, and organizational context (Manning & Cullum-Swan, 1994):

The interpretant, perspective, or standpoint of the observer from which the system is constructed must be identified in social and cultural context. In this sense, a social semiotics requires (or assumes) a rich ethnographic texture within which the semiotic analysis can be socially embedded. (p. 470)

My approach to document analysis is derived from a theoretical and methodological position set forth by George Herbert Mead and Herbert Blumer as well as by Alfred Schutz (1967; see also Berger & Luckmann, 1967) and others. There are three general points that inform the chapters to follow. First, social life consists of a process of communication and interpretation regarding the definition of the situation. The symbolic order we join as infants infuses our own view of ourself, others, and our future.

Second, it is this communicative process that breaks the distinction between subject and object, between internal and external, and joins them in the situation that we experience and take for granted. Our activities are part of the social world we study and are "reflexive" or oriented in the past to what has gone before as part of the relevant process (Hammersley & Atkinson, 1983, pp. 14ff). We try to be aware of this process by being reflective of the overall process, although this orientation could also be said to be reflexive of a theoretical orientation, including assumptions about science and order. Third, the notion of process is key because everything is, so to speak, under construction, even our most firmly held beliefs, values, and personal commitments. What we consciously believe and do is tied to many aspects of "reality maintenance," of which we are less aware, that we have made part of our routine "stock of knowledge."

This is also true of the research process in that it takes place in a historical-cultural context. For example, all research is a social activity. For another, research methods develop in and are influenced by a social context. Finally, research methods and data are derived from a theoretical position about how the world (reality) operates. This means that one selects a method to study certain kinds of data to answer a particular question because there is "reason" (or theory) to expect that the method is appropriate for the data and question in mind. Consequently, a faculty member probably would not ask for students in a classroom to raise their hands if they thought he or she was a poor instructor! (At least, there may be some question about "validity"!) Moreover, as ideas about how the world operates change, so too do our approaches to studying the social world.

In broad terms, these assumptions are consistent with the symbolic interactionists' perspective, which includes a focus on the meaning of activity, the situation in which it emerges, and the importance of interaction for the communication process. Several important concepts relevant to these considerations are context, process, and emergence. *Context,* or the social situations surrounding the document in question, must be understood to grasp the significance of the document itself, even independently of the content in the document. In this sense, archaeologists spend lifetimes trying to understand the nature and cultural meaning of an artifact, some of which any child of the group could explain to them.

Studying documents of our contemporaries or our own lifetimes makes the problem a bit more manageable. For example, one of the most important things to understand about newspapers and TV newscasts is that they are organizational products. This again suggests the importance of *process,* or how something is actually created and put together. Newspapers and TV newscasts are put together according to a routine and a complex division of labor and deadlines. For example, due primarily to entertainment values, TV news formats are oriented to visually exciting and dramatic events, but these can only be filmed (taped) if a crew can know about them before the event occurs or can get rapid access to them (e.g., taping a drug bust). Coverage of events has to be scheduled according to the availability of film crews. Not much detail is necessary. Few local reports will last longer than 90 seconds because TV news organizations assume that few viewers want details or indeed have an attention span to warrant more information. So news reports are very brief.

This general awareness greatly informs the nature of the work that must be completed before a TV report or a newspaper can ever be studied by a researcher. Ideally, any researcher investigating documents such as newspapers or TV newscasts would be familiar with the context and process of each to adequately consider the relevant aspects of a news report, including the knowledge to rule out erroneous explanations of news content. For example, numerous studies of local and national media essentially rule out the importance of a reporter's personal bias in reporting news; this does not mean that there are not exceptions, but for the most part—and quite contrary to many views held by both the left and the right critics of news content—personal bias plays a relatively minor role in shaping news (see Appendix).

Context and process are also important for the meaning and message of a document. These meanings and patterns seldom appear all at once, however; rather, they *emerge* or become more clear through constant comparison and investigation of documents over a period of time. It is because documents provide another way to focus on yet another consideration of social life—emergence—that they are helpful in understanding the process of social life. *Emergence* refers to the gradual shaping of meaning through understanding and interpretation. We use documents to help us understand the process and meaning of social activities. This is very significant in organizations for workers who can use documents as a resource. It is also significant for theoretically informed social analysts, who understand that what people do and how they behave is influenced (but not determined) by their definition of the situation. Symbolic meanings are very important in this process. As we understand the processes used in the production of news or public accounts of social order and disorder, we are able to clarify why the news is what it is and that news content may be as much influenced by the organization of the news process as it is by the events it claims to be about (see Appendix). This is very important: Is crime rampant or is crime covered constantly because of certain organizational and ideological orientations of news producers? Fortunately, numerous studies of the news process have clarified some of these issues, and they are reflected in suggestions for how to study news reports in the following chapters.

It is the interaction between the reader, viewer, or listener of a news report and the news report itself that is important. This may include the situation in which the report is read, seen, or heard—for example, riding a commuter train or sitting alone in one's own living room watching television. The audience member brings experience (context), interest, and degree of awareness to the report, and the complex interaction of these provides the meaning of the report and, therefore, its significance to the individual. The same news report (e.g., a tax increase) will draw different reactions from various readers, and, indeed, the report itself may be interpreted much differently. It is the message in interaction with the audience member (who has a history and context of experience) that provides the meaning of the report for a particular individual.

But why, then, study documents? If one is not interested in the immediate impact (as a bullet) on an audience member, why waste time in carefully investigating documents such as news reports? This is a very important question that requires a complex answer. As noted previously, our perspective on documents has been quite limited by narrow approaches to method and data. The focus on many of the earlier approaches was the individual actor. This remains important but in more recent years, the view has been greatly expanded to include the "effective environment" or the everyday physical and symbolic reality in which people live. Broadly conceived as "cultural studies," this approach seeks to examine the complex interaction between individual perspectives and patterns of meaning and symbolic ordering to understand new sources of social definitions and sort out their consequences. The mass media and the rest of the popular culture industries, for example, are now worldwide, just as our growing awareness of the penetration of information technologies into everyday life and discourse has sensitized us to the power of symbol systems, reality industries, and a host of new realms referred to as cyberspace, hyperreality, and the like. What we want to know is, How do these influence social definitions and social lives? An expanding array of potential documents to investigate has provided new understandings associated with intellectual developments, including critical theory, cultural studies, and postmodernism. Research suggests, for example, that the information technologies and logics guide the content of many aspects of culture in an expanding ecology of communication that views activities (such as news presentation) as information technologies and communication formats that govern the activity. Our awareness of what may be called *hyphenated activities* (e.g., "TV-entertainment-spectacle-news") has led us to a new appreciation of documents as texts that are interpreted and used by audience members.

The nature of the "text process," from initial production to its internal organization, to the "use" and meaning for an individual, are but a few of the challenges for students of culture. We all grow fond of metaphors, but the problem with them is that through repetition they become much more real and taken for granted. For Forrest Gump, "Life is like a box of chocolates"; for some analysts, "Life is like a text." The suggestion is that the metaphor text, which is enacted when an individual interprets a symbolic representation in a specific situation, can also be problematic. *Text,* a term borrowed from literary criticism, is a helpful metaphor, but it can also be limiting if misused and universally applied by fiat to all of social life. Many analysts hungry for a new conceptual handle for complex phenomena have grasped text, but it, as with the positivists' objective reality, will crumble as our insights into even better and richer analytic modes are discovered. In this sense, traveling through culture and hyperspace today is a discovery process for ourselves. Emergence is all we have for certain, but this means that any author's attempt (my own included!) to chart a course through the territory is probably doomed less by poor navigation than by the changing nature of what we are traveling over, through, or in. After all, cyberspace and hyperreality did not exist a few years ago.

These considerations inform an approach to documents. Interest is not primarily in the immediate impact of messages on some audience member, but rather two aspects of the document: (a) the document process, context, and significance and (b) how the document helps define the situation and clarify meaning for the audience member.

Some guidelines for understanding the context of discovery in documents, the process or life cycle of a document, and the use and meaning of that document are presented in Figure 1.1. Although this process will be revisited in Chapter 3, the general flow moves from an original idea about a topic to some ethnographic materials about a relevant or related setting, context, or culture. Step 3 entails actually examining a few relevant documents with this awareness in mind and then, following steps 4 through 12, drafting a protocol for data collection, coding, and analysis and drafting the report.

Documents, then, enable us to (a) place symbolic meaning in context, (b) track the process of its creation and influence on social definitions, (c) let our understanding emerge through detailed investigation, and (d) if we desire, use our understanding from the study of documents to change some social activities, including the production of certain documents!

The following chapters are organized to provide some procedures and rules of thumb for conducting document analysis. Materials from my own or students' projects involving primarily print and television materials will be used to illustrate the steps of this research approach. Following next chapter's comparison of traditional quantitative with ethnographic (or qualitative) content analysis, Chapter 3 provides an overview of the steps in qualitative document analysis and the role of the research task and perspective for various sampling strategies. Chapter 4 incorporates some basic information about information bases in developing protocols and data collection for printed documents. Extending this approach to television and other visual and electronic documents is the focus of Chapter 5, whereas Chapter 6 provides a brief illustration of the approach and advantages of using computer information bases for tracking discourse. Chapter 7 exam-

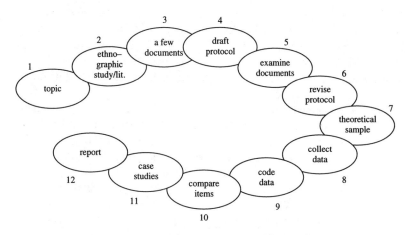

Figure 1.1. The Process of Qualitative Document Analysis

ines field notes and other research materials as documents that can be analyzed more powerfully using the document approach discussed throughout this book.

2. ETHNOGRAPHIC CONTENT ANALYSIS

Introduction

This chapter addresses how an ethnographic approach to document analysis is based on principles of qualitative data collection and analysis. Ethnographic content analysis (ECA) is briefly contrasted with conventional modes of quantitative content analysis to illustrate the usefulness of constant comparison for discovering emergent patterns, emphases, and themes in an analysis of TV news coverage of the Iranian hostage situation. Although later chapters examine how the protocols are constructed to aid in qualitative data collection and analysis, this overview is intended to contrast this approach with conventional quantitatively oriented content analysis. It is suggested that an ethnographic perspective can help delineate patterns of human action when document analysis is conceptualized as

AUTHOR'S NOTE: This chapter is a revision of a 1987 draft of a paper published as "Ethnographic Content Analysis," Qualitative Sociology, *10(1), 65-77.*

fieldwork. Prior research and awareness of an activity involved in the production of documents can theoretically inform sampling procedures, whereas constant comparison and discovery may be used to further delineate specific categories, as well as narrative description. In general, this means that the situations, settings, styles, images, meanings, and nuances are key topics of attention.

It has been claimed that all research directly or indirectly involves participant observation in the selection of a topic, method of study, data collections, analysis, and interpretation (Cicourel, 1964; Hammersley & Atkinson, 1983; Johnson, 1975). Although it may seem evident that any sustained inquiry is constituted through a complex and reflexive interaction process, it is also apparent that some research methods (e.g., ethnography) embrace this process, whereas others (e.g., survey research and content analysis) disavow it. In what follows, I suggest that several aspects of an ethnographic research approach can be applied to content analysis to produce ethnographic content analysis, which may be defined as the reflexive analysis of documents (Plummer, 1983). ECA has been less widely recognized as a distinctive method, although various facets of this approach are apparent in document analyses by historians, literary scholars, and social scientists (Berg, 1989; Glaser & Strauss, 1967; Plummer, 1983; Starosta, 1984). A brief comparison of this approach with conventional content analysis will be preceded by examples of the use of ECA in several research projects.

Ethnography in Context

In general, ethnography refers to the description of people and their culture (Denzin & Lincoln, 1994; Schwartz & Jacobs, 1979). In this sense, the subject matter—human beings engaged in meaningful behavior—guide the mode of inquiry and orientation of the investigator. If the meaning of an activity remains paramount, however, ethnography can also be considered as a methodological orientation independent of a specific subject matter. Products of social interaction, for example, can also be studied reflexively, looking at one feature in the context of what is understood about other features, allowing for the constant comparison suggested by Glaser and Strauss (1967).

An Overview of Content Analysis

Ethnographic content analysis, or what in the remainder of this book I will refer to as qualitative document analysis, may be contrasted with conventional or more quantitative content analysis (QCA), in approach to

TABLE 2.1 Quantitative (QCA) and Ethnographic (ECA) Content Analysis

	QCA	*ECA*
Research goal	Verification	Discovery; verification
Reflexive research design	Seldom	Always
Emphasis	Reliability	Validity
Progression from data collection, analysis, interpretation	Serial	Reflective; circular
Primary researcher involvement	Data analysis and interpretation	All phases
Sample	Random or stratified	Purposive and theoretical
Prestructured categories	All	Some
Training required to collect data	Little	Substantial
Type of data	Numbers	Numbers; narrative
Data entry points	Once	Multiple
Narrative description and comments	Seldom	Always
Concepts emerge during research	Seldom	Always
Data analysis	Statistical	Textual; statistical
Data presentation	Tables	Tables and text

data collection, data analysis, and interpretation. Table 2.1 provides an overview of these approaches on several dimensions.

Quantitative Content Analysis

Originating in positivistic assumptions about objectivity, QCA provided a way of obtaining data to measure the frequency and extent, if not the meaning, of messages (Berelson, 1966). Following a serial progression of category construction-sampling-data collection-data analysis-data coding-interpretation, QCA analysis has been used as a method to determine the objective content of messages of written and electronic documents (e.g., TV cartoons) by collecting quantitative data about predefined and usually precoded categories or variables (McCormack, 1982). This is summarized by Starosta (1984) in the following:

> Content analysis translates frequency of occurrence of certain symbols into summary judgments and comparisons of content of the discourse . . . whatever "means" will presumably take up space and/or time; hence, the greater that space and/or time, the greater the meaning's significance. (p. 185)

For this reason, units of space have most commonly been designated as countable and, therefore, measurable. Even though some of the early proponents of QCA made it clear that no imputation of the speaker's (writer's) motive was warranted, the method was used descriptively as well as evaluatively to relate a message to the source's intention (Berelson, 1966).

The major tact of QCA is to verify or confirm hypothesized relationships rather than discover new or emergent patterns. Indeed, the protocols are usually constructed through operational definitions of concepts to obtain enumerative data for purposes of measurement (Krippendorf, 1978). As this mode of document analysis became further influenced by electronic data processing formats, the primary researcher's role was reduced to setting up the protocol and then analyzing and interpreting data. The data collection and organization (coding) process was carried out by novices hired and quickly trained to find, record, and count the "mentions" for each unit of analysis. Measures of "intercoder reliability" would, presumably, show that the research novitiates were really using the same judgmental criteria in their selection and enumeration. The upshot was that reliability produced validity. Indeed, it was this rationale that led to the institution-alization of intercoder reliability scores on most content analysis studies.

Ethnographic Content Analysis

Ethnographic content analysis is also oriented to documenting and understanding the communication of meaning, as well as verifying theo-retical relationships. A major difference, however, is the reflexive and highly interactive nature of the investigator, concepts, data collection, and analysis. Unlike in QCA, in which the protocol is the instrument, the investigator is continually central in ECA, although protocols may be used in later phases of the research. As with all ethnographic research, the meaning of a message is assumed to be reflected in various modes of information exchange, format, rhythm, and style—for example, the aural and visual as well as the contextual nuances of the report itself.

ECA follows a recursive and reflexive movement between concept development-sampling-data, collection-data, coding-data, and analysis-interpretation. The aim is to be systematic and analytic but not rigid. Categories and variables initially guide the study, but others are allowed and expected to emerge throughout the study, including an orientation toward *constant discovery* and *constant comparison* of relevant situations, settings, styles, images, meanings, and nuances (Berg, 1989; Glaser & Strauss, 1967). To this end, ECA involves focusing on and collecting numerical and narrative data rather than following the positivist convention of QCA of forcing the latter into predefined categories of the former. ECA is oriented to check and supplement as well as supplant prior theoretical claims. The emphasis is on simultaneously obtaining categorical and unique data for every case studied (e.g., news reports about the Iranian hostage crisis) to develop analytical constructs appropriate to several investigations (Schwartz & Jacobs, 1979). As the remaining chapters indicate, data are often conceptually coded so that one item may be relevant for several purposes. In short, although items and topics can still be counted and put in emergent categories, it is also important to provide good descriptive information.

Despite the clear similarities with the approach of *grounded theory,* ECA differs in emphasis and approach. Grounded theory stresses more the systematic coding of field notes, whereas ECA is more oriented to concept development, data collection, and emergent data analysis. The assumption behind ECA is that the general process of data collection, reflection, and protocol refinement is more significant for a study and that details involving coding procedures, practices, and categories do emerge. The major difference, however, is that their foci are different. Grounded theory is trying to generate clear testable hypotheses as a foundation for "theory," and this may require excluding certain materials. ECA is not oriented to theory development but is more comfortable with clear descriptions and definitions compatible with the materials. Central to both, however, is the importance of constant comparison, contrasts, and theoretical sampling.

The Appearance of Crisis

The relevance of reflexive observation can be illustrated by a study of network news coverage of the Iranian hostage crisis, which involved some 52 Americans being held for 444 days—November 1979 to January 1981 (Altheide, 1981, 1982, 1985a). Theoretical and saturation sampling were combined (these will be explained in Chapter 3). Ultimately, hundreds of hours were spent viewing 925 news reports about this highly publicized series of events.

Ethnographers approach a topic with a wealth of information and understanding about human behavior. As suggested in Table 2.1, previous work on TV news had also provided a foundation in news procedures and perspectives that would inform some basic assumptions and approaches in this study, although, in a sense, this was the first study of its kind in that there had not previously been an extended crisis that was so heavily televised, including the Vietnam War. (The closest approximation to the Iranian situation was the imprisonment of the 82 crew members of the

U.S.S. *Pueblo* by North Korea for 10 months in 1968.) This background and related conceptual refinements guided the research task. The task then, was to describe the news coverage but in a theoretically informed manner, which in turn would provide the data necessary for further conceptual refinement.

A Theoretical Focus

The major focus of the study was on concepts and relationships about TV news coverage of an international crisis, particularly the role of formats, including visual imagery, its origins, and relevance for thematic emphasis (Altheide, 1976; Bennett, 1988; Epstein, 1973; Tuchman, 1978). Formats are organizational devices to facilitate coordination of the news process (Altheide, 1985c; Altheide & Snow, 1991). Format refers to the rules and procedures for defining, recognizing, selecting, organizing, and presenting information as news. Communication formats provide an order, logic, and organizationally appropriate template for selecting and shaping relevant material. They are an organizational link to the external environment but also a "probe" of the external environment to find materials that are compatible with organizational criteria. The dimensions of TV news formats include short reports with visual and aural information that can be presented in a narrative form with a beginning, middle, and end. It is possible to adapt almost any event to this format, but events with certain characteristics are more likely to be selected for coverage because they can more readily be shaped according to the logic of format (Altheide & Snow, 1979, 1991). Event characteristics most relevant include accessibility, visual quality, drama and action, perceived audience relevance, and encapsulation and thematic unity. The general dimensions of each can be briefly stated as follows:

> *Accessibility* refers to how easily newsworkers can learn about an event, obtain information about it, get to a site where it occurs, and/or obtain visuals.
>
> *Visual quality* is the extent and clarity of film, tape, or other visual depictions of the significant action.
>
> *Drama and action* refer to the graphic, visual, and aural portrayal of some movement that is used to illustrate the event.
>
> *Encapsulation and thematic unity* refer to the ease with which an event can be (a) briefly stated and summarized and (b) joined to a similar event or a series of reports over a period of time or even within the same newscast.

Audience relevance is the interest an item is perceived by newsworkers to have for a mass audience.

Of course, the extent of the audience's interest is also a feature of the way the other format criteria are brought to bear on an event. This is often assessed by producers' assumptions of sharing a common culture with their audience—that is, "we'd be interested in this"—but also includes awareness of what other popular culture messages have been presented. If a report reflects "common culture" assumptions (e.g., common fears), and other media outlets have presented similar reports, then it is "obvious" to any competent member of the organization seeking to provide viewers with "what they want" that they should participate in what by then has become "what all America is talking about."

Sampling and Data Collection

Data were obtained through content analysis of selected newscasts in two waves: November 4, 1979, to June 7, 1980, and July 3, 1980, to January 24, 1981. (Intercoder reliability on a sample of 36 newscasts was 91.7%!) Prior to systematically selecting newscasts that would adequately represent this more than 14 months of daily coverage, all available news reports pertaining to Iran on the network evening newscasts during the first 9 days of the embassy takeover (November 4-12, 1979) were viewed and analyzed. This preliminary analysis made it clear that a simple random sample or even a stratified sample would systematically distort the uneven coverage provided to some aspects of this major news event, including concentrations around holidays as well as the tendency to broadcast a certain feature of the ordeal over a period of several days, a common feature of thematic emphasis. For example, seldom would a simple random sample select 2 or 3 days in succession, although the networks frequently stretched a series of reports over several days (e.g., the reaction to Iranian students). This strategy, along with checking news records from the Vanderbilt University Television News Archive, provided qualitative data as well as enumerative data that indicated that the networks were quite similar in the amount and emphasis of their coverage (Altheide, 1981). This archive text is now available on an Internet "gopher" and will probably have visuals available for "downloading" within a short time.

To obtain a theoretically informed sample while still drawing a comparative sample of news reports about Iran, a saturation sample was drawn covering approximately 112 days and 26 hours of compiled news reports.

This was done by combining two units of analysis: each network report pertaining to Iran (which may have included more than one topic) presented on a given newscast and several consecutive newscasts, or "clusters." Overall, 17 clusters were drawn consisting of 5 to 9 consecutive newscasts per network. Care was taken to proportionately represent all 14 months, as well as weeks 1 through 4 of the various months.

The original data collection protocol was constructed to provide both numeric and narrative (descriptive) data collection for (a) network; (b) presenter; (c) length of report; (d) origin of report; (e) news sources; (f) names and status of individuals presented or interviewed; (g) their dress, appearance, and facility with English; (h) what was filmed; and (i) the correspondence between film, speech, and overall emphasis. When additional observations offered additional subtopics, the narrative portion was particularly helpful for reexamination and for developing a framework for dealing with *visuals.*

With the exception of identifying materials (e.g., network, date, time, etc.), the content was viewed with an emergent orientation to minimize predefined and rigid categories for defining what was relevant. This approach produced the five "event characteristics" of TV news noted above. The general procedure was to view a few reports, assess the message(s) for news techniques, and, only then, record some general categories for this report as well as for reexamining several previous reports. It was then important to check the quality and quantity of information being recorded involving such queries as, What is being omitted? or, What segments, time blocks, and so forth, do not seem important for the present focus?

This constantly refined exploration and comparison provided a substantively informed sampling procedure, as well as the category of *topical emphasis* to guide data collection. The main topics were defined as follows:

Hostages: Any report that focused on the hostages' status, location, health, and so forth.

Families: Reports focusing on the hostage families' status, health, reaction, plans, and so forth.

Shah of Iran: Reports dealing with the context of the Shah's rule, including political alliances, enemies, as well as his status, health, location, and statements.

Iran: Iranian government action, plans, reactions, statements, and elections.

Iran (internal problems): Reports pertaining to economic, civil, criminal, and demonstration problems, such as internal revolts.

Iran (external problems): Reports involving economic sanctions or military threats, such as the Soviets in Afghanistan and the war with Iraq.

USA: Reports involving U.S. government actions, statements, reactions, proposals, criticisms, and so forth.

International: Those reports involving international statements, actions, reactions, proposals, involvements, sanctions, including the United Nations and the World Court.

Iranian residents: Reports about Iranian students in the United States, reactions to U.S. policy, support of the Iranian government, demonstrations, civil and criminal actions, and so forth.

The progression from data collection to interpretation mirrors the discussion in the remaining chapters. It was intended to be reflexive and nonlinear (e.g., constant discovery) because previous research with numerous journalists revealed that newsworkers often incorporate file film and old reports into new ones (Altheide, 1976) so that any update or overview of a report is usually tied to what has gone before. In this sense, TV news often reports on itself.

A few brief examples illustrate what was gained from viewing news content reflexively rather than statically. If the QCA approach to sampling and data collection had been followed, important thematic patterns would have been lost. Major reports about various facets of the hostage situation occurred in miniseries, often over a period of several days. The significance and message of one report would be lost when removed from the context of other reports. For example, although the families of hostages were featured in about 12% of the total sample of 925 news reports, they were involved in 37% of all hostage-related reports during the hostages' first Christmas and 25% of reports about the second Christmas. The upshot is that the hostages' families played more of a role at certain times, and this role had a great deal to do with the format of TV news.

One challenge for this lengthy study was to articulate the relevance of visuals within the news format. Ethnographic content analysis offers a perspective for systematically studying the use of visuals and text in media reports (Adams & Schreibman, 1978; Lang & Lang, 1968; Tuchman, 1978). Comparative study of the thematic and visual coverage of the topics noted previously provided a way to empirically ground a theoretical perspective about the role of formats in news definition, selection, organization, and presentation. For example, the news sources and visual opportunities fluctuated during the hostage ordeal so that new visuals of hostages and their captors became quite scarce. The networks' problem, then, was to look elsewhere, but where they looked also led to different topics (e.g., family members), emphases, and conclusions. An atheoretical sample would have missed these systematic clusters that show very clearly that the substance of reports could be predicted by the origin of reports (Altheide, 1985c, p. 77ff).

Data Analysis

The realization that the origin of reports and visuals were joined through thematic emphasis did not emerge until well into the study. My procedure was to narratively describe the news visuals by "what was shown," "who was shown," and "what they were doing." Rereading my descriptions and relating these to certain events (e.g., the hostages' Christmas) led me to return to previous tapes and add additional data to the open-ended protocols. This in turn led me to realize that "file tape" was more often used with reports originating from certain locations, which in turn were more likely to be associated with certain topics and aspects of the hostage situation. For example, visuals—usually file film—of crowds chanting anti-American slogans were routinely used when a reporter in London would conclude a summary of the day's nondevelopments. This mode of analysis permits emergent conceptual relationships to unfold that would not be discovered from cross-tabulating two or more precoded variables.

The analysis of TV visuals (discussed in Chapter 5) benefited from the exploration and experimentation with several categories and codes in this study. A qualitative document analysis seeks to illustrate relevant categories. For example, when studying visual documents, one needs a way to delineate what is actually stressed in news reports. If the hostages' families were symbolic of the American perspective, visuals of Iranian crowds chanting anti-American slogans emerged as the visual definition of the adversaries. For example, there were 87 film reports (9.4% of total) about Iranians, primarily in street demonstrations and crowd scenes. Iranians in the United States were similarly presented: 31 of 64 reports featured them in crowd activities, including broadcasts of some of the most brutal confrontations since the Civil Rights movement. A related concern is how a reporter's narrative, with its emphasis and theme, is visually presented on TV. What does "conflict" and "normal" life look like? One consequence of the visual emphasis was to further compound an already strange encounter with adversaries from a different culture, with a different worldview and a very different religion. For example, on January 16, 1980, the following exchange took place between an anchorperson and a reporter in Iran.

Anchor: What is happening there? Do Americans really understand what is happening there?

Reporter: I think . . . that the impression we convey from the scenes in front of the embassy, all the fist shakers yelling "Death to Carter, Death to America," we conveyed a picture of a nation in the grip of madness, and yet just a few blocks away from the embassy gates people are going about their lives in a normal fashion. Mothers are taking their babies to the park. Businesses are opened. Tehran is pretty much working as normal (NBC Special Report, 1980).

There is still another important reason for this ethnographic approach. If one uses only structured protocols in content analysis research, critical questions, issues, and shortcomings that may become apparent at a later time may have to be forsaken due to lack of data. If, however, ethnographic materials are also included, then it is usually possible to return to the data when other questions and inquiries arise. I have been able to use old data sets for more current reflections because of this. It is the way this process works that guides the remaining chapters.

3. PROCESS OF
QUALITATIVE DOCUMENT ANALYSIS

The comparison in Chapter 2 between conventional quantitative content analysis and ethnographic content analysis noted several important differences. Several of these will be discussed in more detail following, particularly description and emergence. Figure 1.1 in Chapter 1 illustrated 12 steps in document analysis that essentially encompasses several parts: finding and gaining access to the documents, collecting data from them, organizing the data, and analyzing the data. Figure 3.1 illustrates the twelve steps as a process involving five stages: (a) documents, (b) protocol development and data collection, (c) data coding and organization, (d) data analysis, and (e) report. Icons are used in this figure and several others to reflect different document sources, ranging from TV, video, film materials, books and manuscripts, files and notes, and photographs and other visual records. The diskettes represent computer-assisted database managers or text files.

The Problem and the Unit of Analysis

Step 1. Pursue a specific problem to be investigated.

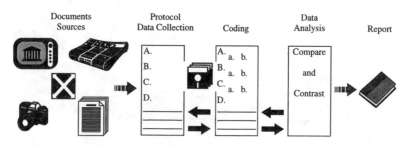

Figure 3.1. Stages of Document Analysis

Step 2. Become familiar with the process and context of the information source (e.g., ethnographic studies of newspapers or television stations). Explore possible sources (perhaps documents) of information.

Step 3. Become familiar with several (6 to 10) examples of relevant documents, noting particularly the format. Select a unit of analysis (e.g., each article), which may change.

The research problem helps inform the appropriate unit of analysis, or which portion or segment of relevant documents will actually be investigated. When studying the news coverage of the Iranian hostage crisis, I had to decide whether the entire newscast should be examined or if the individual news reports should be the unit of analysis. As noted in Chapter 2, qualitative document analysis relies on the researcher's interaction and involvement with documents selected for their relevance to a research topic. Although most research questions begin with notions such as "why" or "what is the impact of . . . ," they tend to be quite abstract and far removed from available research materials. My approach is to go ahead and ask those questions (e.g., "Why did the networks present the Iranian hostage crisis a certain way?") but to then break it down into several parts, including, "*How* was the Iranian hostage crisis presented as TV news reports?" Implicit in this approach is to combine "how" with "what," as in "what was said and shown?" More specifically, include categories to document the act, scene, agent, agency, and purpose reflected in the document (Brissett & Edgley, 1990; Burke, 1962).

These questions were consistent with my familiarity with the nature and process of TV news work, including all aspects of the production process— for example, the important role of news sources that provide items that can fit with the news workers' schedule and preference for zippy visual reports. TV news, after all, is primarily entertainment oriented, and the major news sources have recognized this for some time. This is the ideal experience and awareness of context that is useful to have before beginning a study of TV news, newspapers, or advertisements, for example. Few people, however, have this experience, but they can still do a first-rate job of document analysis. A substitute is to read one or two accounts of the document-production process—studies of news work, for example—to become more familiar with the rudimentary process of producing the information with which you will work. There are numerous accounts available in journal and book form about the production process of newspapers, TV news, entertainment TV shows (e.g., soap operas), and so forth. The same is true of the use of other documents (e.g., the FBI Uniform Crime Reports or Academy Award nominations). Try to be familiar enough with the process for collecting or creating the information to be used that you can describe the key steps in its production.

Familiarity with TV news production indicates that TV operates with time, meaning that it allocates portions of its newscast to certain topics and that the ones that receive the most air time are usually those regarded as the most important. If these reports also come early or at the "top" of the newscast, they are very important. Consequently, one component of the general question of "what" or "how" did TV news cover the Iranian hostage crisis boils down to how much time was allocated to such coverage for each newscast.

The problem to be investigated helps clarify the unit or level of analysis. For example, as noted above, in TV news reports, one could use the entire newscast, individual news reports, or parts of reports. On examining, recording and reviewing, and thinking about several news reports of the Iranian hostage crisis, it became apparent that although there were different parts to individual news reports, the individual reports would be the unit of analysis. If one is studying newspapers, the unit of analysis could be a particular page, individual articles, or perhaps even paragraphs. In most cases, the unit of analysis is the individual news report. In studies of television violence, the unit of analysis could be a week, an evening, a program such as certain cartoon shows or individual acts presented in a program.

Constructing a Protocol

Step 4. List several items or categories (variables) to guide data collection and draft a protocol (data collection sheet).

26

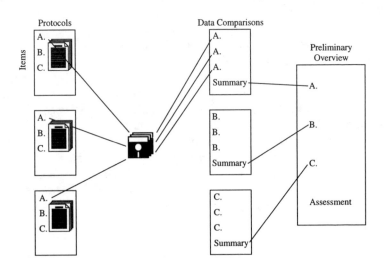

Figure 3.2. Logic of Protocol Analysis

Step 5. Test the protocol by collecting data from several documents.

Step 6. Revise the protocol and select several additional cases to further refine the protocol.

Figure 3.2 illustrates the logic of protocol analysis: One moves from data collection on various items, categories, or variables (diskette) to data comparison and summary within each category with the aim of writing preliminary statements about category findings. All of this begins with the protocol.

In general terms, a protocol is a way to ask questions of a document; a protocol is a list of questions, items, categories, or variables that guide data collection from documents. We want to ask the right questions—that is, those that are conceptually cogent. In quantitative content analysis, the emphasis is on obtaining data that can be counted and analyzed statistically. Protocols for quantitative content analysis tend to have numerous categories or variables, often numbering in the hundreds. Usually, quantitative content analysis protocols are precoded before the data are collected. The main emphasis of qualitative content analysis is on capturing definitions, meanings, process, and types. Although these can be listed numerically (e.g., how many news sources and with what frequency in reporting the Iranian hostage crisis), these are used to supplement understanding and interpretation derived from other data as well. Consequently, qualitative document analysis relies a good deal on text, narrative, and descriptions. For this reason, protocols for qualitative document analysis tend to be less precise and fairly short, often having a dozen or fewer categories. These protocols may have some precoded items for each of the categories, but most are likely to be coded and given "refined meaning" after the data have been collected.

The rules of thumb are quite simple. First, treat the development of your protocol as a part of the research project and let it emerge over several drafts. The ultimate test of a protocol can be carried out with a few cases simply by asking, "Can I obtain the necessary information for my study from the cases (e.g., TV newscasts or newspaper reports) using this protocol?" If not, amend your protocol, and if this seems too muddled, consider amending your list of research questions. Second, keep categories to a minimum at first, but others can be added as the investigator interacts with documents and relevant theoretical issues. Third, no item in the protocol should stand alone or be included just because the answers would be interesting. Rather, all items included in your protocol should be relevant for at least one other item. Fourth, your protocol should be capable of accommodating both numerical or letter codes as well as descriptions. Your best material will come from descriptions and even quotations from the document. Fifth, in general, your protocol categories should have more than one possible outcome or value to them. For example, if you are using different newspapers in your study, one item would be Newspapers, with choices a, b, c, and so forth. Sixth, protocols for qualitative document analysis in most cases should include categories that are relevant for characteristics of social action, including providing information about the time, place, and manner of an activity. Stated differently, What or how is it done? Where and when was it done? Who did it? With what rationale? Were any motives apparent? The goal is to show that the document reflects social activity, and the categories noted above have been shown to be useful ways to capture the "dramaturgical character" of action. Seventh, if appropriate, protocols of documents should be capable of documenting visual as well as written and narrative information. Eighth, protocols should have a reflective segment in which the researcher can make research notes and comments about how this case was similar to or different from others.

Several items pertinent to Point 6 also illustrate some additional features of a protocol. Although there will be differences due to the subject matter and the source of documents under investigations, there are some generic categories that can be included in most protocols. First, there should be categories for the following: case number, medium (e.g., *New York Times, Los Angeles Times,* etc.), date, location, length (space or time), title or emphasis, focus or main topic, source(s), and themes. If news materials are being analyzed, there should also be a category relevant for format considerations, particularly visuals for television (e.g., types of visual(s)—videotape, file tape, graphics, and about who, and what are they doing). There should also be a section for a brief description in your own words. There may be subcategories for each. This bare bones outline of the major categories will probably be supplemented with several additional items for the researcher's specific interest (e.g., crime reports—type of crime).

The protocol will be edited and revised as suggested in Steps 5 and 6 from Figure 3.1. The test of a good protocol is if your conceptual problem is adequately covered by your categories. One additional way to check this is to reread a document from which data has just been collected. Ask yourself, point by point, if the thrust or essence of the article for purposes of your study has been adequately captured by the protocol data. If not, ask if an additional item or a revision of an existing one would improve the fit, and if so, make appropriate adjustments.

When the protocol is adequate, additional documents may be selected for data collection. (Later chapters will address different ways and sources to select documents.) The best way to record data seems to vary with the researcher. Perhaps the easiest is to make copies of the protocol and attach one copy to each document providing data. These hard copies can be stapled together and filed appropriately.

A simple way to use your protocol is to have a blank copy of the protocol in a file and then copy it for additional data documents. These should be backed up and saved on a hard drive as well as a disk or two. Based on my experience as well as that of several hundred students I have supervised, I would say that the rule of thumb is that at least one (or two) of these backups will fail during the course of your study, so the inexpensive disks are cheap when compared to losing data (e.g., one student's dog ate her disk!).

Themes and Frames

A key category in most qualitative studies of documents involves meaning and emphasis. Several overlapping concepts that aim to capture the emphasis and meaning are frame, theme, and discourse. As anyone who has tried to explain these to students realizes, however, they are much easier

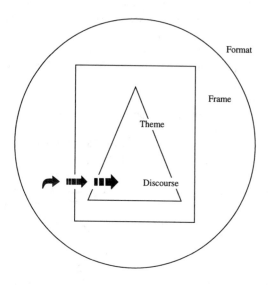

Figure 3.3. Format, Frame, Theme, and Discourse

to "talk about" than to specify for research purposes, particularly pinning them down to the specifics of a protocol so that you know it when you see it! I cannot deliver on the precise details for every study, but I can specify the concepts and show how they are related so that these concepts can be translated into more precise categories for data collection.

Frame, theme, and discourse are also related to *communication formats,* which, in the case of mass media, refer to the selection, organization, and presentation of information. Formats pertain to the underlying organization and assumptions of time (temporal flow and rhythm), space (place and visual editing), and manner (style) of experience. Formats, basically, are what make our familiar experiences familiar and recognizable as one thing rather than another—for example, we can quickly tell the difference between, say, a TV newscast, a sitcom, and a talk show. Figure 3.3 illustrates the relationship of format, frame, theme, and discourse. Our previous work has identified several dimensions of formats that significantly contribute to frames, themes, and discourse (Altheide, 1985c, 1995; Altheide & Snow, 1991).

Communication and media formats enable us to recognize various frames that give a general definition of what is before us. Goffman (1974) referred to *frames* as "schematic of interpretation . . . which enable people to locate, perceive, identify and label 'occurrences of information' " (p. 55). Indeed, one perspective on framing analysis is that it focuses more on the structure of reports (e.g., the ordering of stories or the use of film) than on the content. For example, Zhondang and Kosicki (1993) see framing analysis as the following:

> Framing analysis . . . a constructivist approach to examine news discourse . . . [with a] focus on conceptualizing news texts . . . [as] syntactical, script, thematic, rhetorical structures . . . so that evidence of the news media's framing of issues in news texts may be gathered. (pp. 55-58)

For purposes of this book, however, I will treat frames as very broad thematic emphases or definitions of a report, similar to the border around a picture that separates it from the wall and from other possibilities. An example is treating illegal drug use as a "public health issue" as opposed to a "criminal justice issue." These are two different frames that entail a way of discussing the problem or the kind of discourse that will follow.

Themes are more basically tied to the format used by journalists who have a short time to tell a story that their audience can recognize and that they have probably heard before and, moreover, to get specific information from sources that can be tied to this. That is where sources of information get linked to news media—they not only have the information but they also have learned to put it together in ways that are compatible with the different media formats.

My preference, however, is to think of themes as general meanings or even "miniframes" for a report. This is most apparent in news reports, but it is apparent in numerous other documents as well (Altheide, 1976; Berg, 1989; Epstein, 1973; Fishman, 1980; Zhondang & Kosicki, 1993). Themes are general definitions or interpretive frames (e.g., "Cities are more dangerous than ever," "the most corrupt administration in history," or "modern life is sick." In the case of political coverage, there is the "horse race theme"—which candidate is leading or how public opinion poll data are used to make a "race more exciting"—but there is also the "underdog theme," as when a "front-runner" candidate is overtaken or when an individual sues a corporation—also referred to as a "David and Goliath theme"). Most documents will refer to a theme or point of view, whereas other documents will use "angles" to refer to specific parts of a theme (e.g., the theme of the "most corrupt administration in history" can be supported by the angle of "pressuring congressional representatives to not vote for impeachment").

Obviously, themes and frames are related, but they are not determinate. Different frames can be used within the same theme—for example, punishment and suffering. *Themes* are the recurring typical theses that run through a lot of the reports. *Frames* are the focus, a parameter or boundary, for discussing a particular event. Frames focus on what will be discussed, how it will be discussed, and above all, how it will not be discussed. Certain themes become appropriate if particular frames are adopted. Thus, within the criminal justice frame, which implies a discourse of punishment, themes of health care, including treatment, intervention, and even education, seem a bit out of place.

The upshot is that the actual words and direct messages of documents carry the discourse (more on this in Chapter 6) that reflects certain themes, which in turn are held together and given meaning by a broad frame: Is this a newscast or a standup comedy routine (van Dijk, 1988)? Finally, what media are involved and what are the prevailing format features? For example, the entertainment format has been identified as being much different in electronic visual media, compared to print.

I have found it useful to use the concepts of frame and discourse in studying news reports. As I have stressed, frames are a kind of "super theme." Although there are subtle distinctions, they can be closely joined when examining documents: Frames and discourse help us appreciate this relationship, even though these concepts are also related. Discourse refers to the parameters of relevant meaning that one uses to talk about things. Frame refers to the particular perspective one uses to bracket or mark off something as one thing rather than another. Meaning and language are implicated in both. We can simply say that discourse and frame work together to suggest a taken-for-granted perspective for how one might approach a problem.

The significance of frames, themes, and discourse for document analysis cannot be overemphasized. Theoretically, frames and themes are crucial in defining situations and provide much of the rationale for document analysis. These are the most powerful features of public information, and the study of their origins, how they change over time, and their taken-for-granted use in everyday life is essential to understanding the relevance of communication media for our lives.

Gitlin's (1980) observation about the role of themes and frames in shaping the student protest movement is an apt illustration of the significance for news content. Gitlin notes that the media influenced the movement by drawing on certain frames that helped the activities and meanings, although very selectively: He lists trivialization (making light of movement language, dress, age, style, and goals), polarization (emphasizing counterdemonstrations), emphasis on internal dissension, marginalization (demonstrators as deviants), disparagement by numbers, disparagement of the movement's effectiveness, reliance on statements by government officials and other authorities, emphasis on the carrying of Viet Cong flags, emphasis on violence in demonstrations, de-legitimizing use of quotation marks around terms such as "peace march," considerable attention to right-wing opposition to the movement, especially from the administration and other politicians (in Graber, 1984, pp. 243-244). Gitlin further states the following:

> Some of this framing can be attributed to traditional assumptions in news treatment. . . . Some of the treatment follows from organizational and technical features of news coverage—which in turn are not ideologically neutral. . . . The proportion of a given frame that emanates from each of these sources varies from story to story. . . . (p. 244) The more closely the concerns and values of social movements coincide with the concerns and values of elites in politics and in media, the more likely they are to become incorporated in the prevailing news frames. (p. 247)

Step 7. Arrive at a sampling rationale and strategy—for example, theoretical, opportunistic, cluster, stratified random. (Note that this will usually be theoretical sampling.)

Our understanding about the topic influences our awareness of where we should look for documents. There are numerous sampling strategies, but only a few of the most appropriate for qualitative document analysis will be discussed here (Berg, 1989; Hessler, 1992). The rationale for qualitative research emphasizes clarifying the process, the types (or definitions) of what is presented, and the emphasis and meanings of the messages. This approach also influences the sampling strategy in that the main goal is seldom to "generalize" one's findings to an entire population. To do this, some kind of probability sampling would be necessary, such as simple random sampling or stratified random sampling. Although qualitative document analysis is not mainly concerned with generalization to a larger population, it can easily accommodate this consideration.

An optimum sampling strategy will permit comparisons and contrasts in that "facts" or findings by themselves may be interesting but do not provide conceptual clarity or understanding—as in, "So what?"—unless they can be compared with something else. This must be considered in developing a sampling approach. Two of the easiest ways to do this are to compare documents from different TV networks, different newspapers, or perhaps different countries. Another way is to study different media—print versus television—and different time periods—the start and the end of the "crisis." All of these concerns influence the sampling strategy that is used.

Theoretical Sampling

The major emphasis of qualitative document analysis is to capture the meanings, emphasis, and themes of messages and to understand the organization and process of how they are presented (Glaser & Strauss, 1967). This requires that we include the widest range of relevant messages in our sample. It is difficult, however, to know what this range and variety is at the start of the research. It must emerge as the researcher inspects and reflects on some initial materials. A researcher would be advised to have an "ideal" about the kinds of materials that would be included in the study to answer the major research questions. With these in mind, a "reality" check can then be made against practical limitations such as time, access, availability, and research funds. For example, a researcher might want to investigate the decision-making process of presidents and their advisors. Ideally, this might entail having direct access or even being a participant in behind-the-scenes conversations about certain issues discussed at the White House. This, however, is not likely to be possible, so a "mental search" might ensue for alternative sources of information. One possibility, for example, would be the infamous *White House Transcripts* pertaining to the Watergate investigation and cover-up during the Nixon administration. Even though these tapes and transcripts have been edited, these materials, along with subsequent trial transcripts, are rich documents, some of the most personal, to my knowledge, of a behind-the-scenes look at political power (Molotch & Boden, 1985). Of course, there may be others, but the point is that the investigator can compare what he or she is left with after beginning with an ideal. It is surprising what other sources of information may become available with a little investigation. Indeed, university reference librarians are often a tremendous resource to check out.

Rather than "trap" our analysis with too many preset categories and cases derived from a rigid sampling strategy, it is better to use "progressive theoretical sampling." This refers to the selection of materials based on emerging understanding of the topic under investigation. The idea is to select materials for conceptual or theoretically relevant reasons. For example, a researcher might want to include materials that are similar or different on a particular dimension. Two brief examples can illustrate this. One is taken from newspaper reports, the other from TV news.

In a study of newspaper coverage of police officer shootings, one of my students, Jen Ferguson (1995), was interested in how newspapers report different kinds of police shootings. As a novice student of media processes, Ms. Ferguson discussed, read, and learned about the organizational context of news reports, particularly the importance of news formats, news sources, and organizational routines in covering and presenting public messages. This took some time and effort to digest the context and environment of news reports. In addition to obtaining a background understanding in the news process, the concept of "accounts" and the types—excuses and justifications—guided her inquiry (Scott & Lyman, 1968).

The first step in theoretical sampling for this project was to find some relevant news reports. Using LEXIS/NEXIS, a computerized information base, she searched for several instances of "police shootings." She identified several key "search words" that appear very often in these stories. This initial investigation provided several examples, which she then read several times to develop her protocol. These reports were about instances in which officers shot individuals suspected of criminal activity. Closer reading revealed that there were other types as well, so she conducted additional searches for some different types. Ultimately, she was able to clarify her range of reports to include the following seven types of police shooting:

1. A suspect who shoots at them
2. A suspect who pulls a weapon on them
3. A suspect who pulls a weapon on another individual
4. An armed suspect
5. An individual based on the perception of a criminal activity
6. A bystander
7. Another police officer

Her use of theoretical sampling ensured that the full range of situations was identified.

The significance of this is that a better understanding can be developed of how accounts are used across a range of situations in which police have been involved in shootings. Briefly, accounts refer to statements that are made when routine expectations are violated. In a sense, accounts are "social repair" devices to indicate that although something out of character or place occurred, it can be explained. Accounts can be distinguished as "excuses" and "justifications." With the former, the agent (individual or governmental representative) making the statement acknowledges the act is not really appropriate but denies responsibility for it, such as "I know I bumped into you and that should not happen, but I slipped." With justifications, on the other hand, the agent denies that the act was wrong and accepts responsibility as "I bumped into you, but you were in my way." Adding this conceptual distinction as a data collection and coding category to a study of documents opens up another useful area of inquiry. This helps to assess which situations are likely to be explained through the use of an account that helps to understand how accounts given in the media are influenced by the source that contributes to the account. With examples of each type, she continued to refine her protocol:

> This information was descriptive in nature and involved recording the characteristic of the event that occurred, who was involved in the incident, what the [reported] perceptions of the police were that led to the shooting, who is the source of the account [and whether more than one source was mentioned], what was said about the police and the person [especially "character attributes"] who was shot, what the reason was for the shooting. (Ferguson, 1995, p. 7)

Her rich analysis provided some useful insights, particularly the way different language was used involving some shootings in which the "gun discharged" as opposed to those in which "the officer fired his weapon." Responsibility is implied in the latter and not the former. In addition, this provides still additional search terms to expand the theoretical sample.

This also raises the issue of how many cases are necessary. How many reports should be analyzed to have confidence in the findings? There is not one answer, but it depends on the purpose of the research. In most instances, qualitative document analysis is focusing on the range of meanings and themes as well as process or logic behind reports and emphases. The researcher needs enough reports to demonstrate the range and the differences that make up the range. Ideally, one would have several examples of each to clearly illustrate the process. Thus, an initial investigation into the types, emphases, meanings, and process may require 15 to 20 reports, depending on resources, time, and access to materials. This research, however, is also cumulative, so that this initial "exploratory" work can be added to with a larger sample later. This brings us to a discussion about representativeness.

Stratified Random Sampling

Stratified random sampling refers to random selection of cases within certain categories or strata (Henry, 1990). The strata are selected for conceptual reasons. In qualitative document analysis, stratified random sampling can be used at a later point in the analysis to provide cases to supplement those analyzed using theoretical sampling.

In qualitative document analysis, the frequency and representativeness is not the main issue, conceptual adequacy is. Once the key elements are identified and demonstrated and the data gathering protocol and relevant codes are refined, then the sample can be extended to focus mainly on the entire time period under consideration. Whereas the initial study may have involved 15 to 20 reports, if the event under investigation, such as the Iranian hostage crisis, lasted for months and received hundreds of reports, additional reports may be added to the sample to cover the relevant time frame. Ultimately, my study of the Iranian hostage crisis covered more than 900 reports over 444 days. This many reports, however, would not be necessary now because several conceptual points about sampling were demonstrated and clarified in this and other work.

Selecting additional reports need not be purely random, but it could be, depending on the researcher's conceptual interest. Pure random sampling is very simple: Each day's coverage of the event would have an equal chance of being included in the total population. The researcher would simply note the number of reports that occurred and randomly draw out a predetermined number to have an adequate representation. Virtually any statistics book can be consulted for a formula of how large the sample would be, but a good rule of thumb is 5% to 10%.

If, however, the research has progressed along the lines discussed here, there is a more conceptually informed way to combine theoretical sampling with random sampling. It was mentioned previously that one approach is "cluster sampling" to cover the range of topics and events. If one also wants to cover a particular time period then "stratified random sampling" or "purposive random sampling" is useful. In the case of the Iranian hostage crisis, for example, the extensive involvement and interaction with the reports and the crisis as it unfolded made it clear that there were different stages, steps, or aspects of the event and coverage, including the "pre-" and "post-" ban of foreign journalists.

Collecting the Data

Step 8. Collect the data, using preset codes, if appropriate, and many descriptive examples. Keep the data with the original documents, but also enter data in a computer-text-word processing format for easier search-find and text coding. Midpoint analysis: About halfway to two thirds through the sample, examine the data to permit emergence, refinement, or collapsing of additional categories. Make appropriate adjustments to other data. Complete data collection.

Qualitative document data are very individualistic in the sense that the main investigator is "involved" with the concepts, relevance, processual development of the protocol, and internal logic of the categories, or the way in which the items have been collected for purposes of later analysis. This is particularly true the more exploratory the study is and the fewer categories—and therefore less refined—they are. This does not mean that another investigator or assistant cannot be trained or oriented to collecting the same information or that researchers cannot reach agreement on most aspects of the protocol. It simply means that it will be more problematic and will take considerably more interaction than is the case with conventional quantitative content analysis between the investigators and the protocol, the investigators themselves, and the investigators and the problem under study.

Data are collected by providing relevant codes and descriptions to the protocol categories. Below I present a TV news report from ABC and a completed protocol for a research project on the use of "fear" in news reports. Note the following materials taken from the news report and used in the protocol as illustrations of the emphasis and language.

ABC News
Show: "World News Sunday"
May 3, 1992

Forrest Sawyer: One way or another, people all across the country have been trying to understand just how these past five days could have happened. There are lessons for everyone, including our very young children, most of whom never saw anything like this before. Here's ABC's Karen Burnes.

Karen Burnes: [WATTS FTG] In 1965, angry mobs stormed white police officers who arrested a black motorist in the Watts section of Los Angeles. Their rage and their cries for justice flared into riots. ['92

RIOTS] Last week, the torch was passed to a new generation, who took to the streets of south-central Los Angeles after four white police officers were acquitted in the beating of black motorist Rodney King. [KING VIDEO] They too, demanded justice. It has been nearly 30 years. Racism, say sociologists, has not diminished with time. Now there is this generation, innocent and impressionable, pivotal if history is not to repeat itself. [DENNY VIDEO]

Professor Lillian Katz/U. Illinois: You are not born with a hatred or prejudice toward any particular group. [KIDS IN SCHOOL] It is learned and it's learned easily.

Karen Burnes: [KIDS PLAYING] Children learn prejudice over time and in stages. Experts say they notice differences like skin color and facial features by about age 2. As their exposure increases, so does their awareness.

Professor Lillian Katz: Certainly by 4, most American children are aware of racial characteristics, but that isn't the same as being prejudiced.

Karen Burnes: [OVERTURNING CAR SCENE] Although many young children have already been nurtured on violence, exposed to brutal images, their attitudes have not yet hardened. This is their perception of what happened to Rodney King. [TO BOY] Why did he get beaten up?

Elom Ketosugbo/Boy: Because he was driving too fast.

Sefaka Ketosugbo: Maybe they're in a bad mood and they were in such a bad mood that they just beat him and they forgot the rules, like to give him a ticket or something.

Pat Miller: Alex said, "Well, he looks like such a nice man. Why did anybody beat him up?" He seems really sad about that.

Karen Burnes: [ALEX WITH MOM] By the time children are about 6, they begin to mimic parents and those they see at school.

Darlene Hopson/Psychologist: Oftentimes when children maintain a negative attitude about people who are different from them, it's somehow being encouraged or accepted in the home environment.

Pat Miller: We are their first teachers, so if we have an intense reaction to something, 2 days later you're going to see a child have an intense reaction to that same thing.

Karen Burnes: [FATHER WITH SON OUTSIDE] Prejudice is often simply fear. For a child, it is fear of the unknown.

Professor Lillian Katz: They need exposure to each other so that they can see differences between groups—their own group and other groups— so that they don't stereotype them.

Karen Burnes: And finally, say experts, if children are proud of their own culture, they'll be more comfortable and tolerant of others.

Sefaka Ketosugbo: Everybody should be equal in this world.

Elom Ketosugbo: They should be free and not be hurt anymore.

Pat Miller: You know, the kids are the hope of the future, and if they are brought up to think that it's wonderful that there are different races and colors and creeds in the world and that it's absolutely normal and the way it was meant to be, then maybe then when they're the politicians and the people that are discussing these issues, we'll have a better chance. [CU KIDS FACES]

Karen Burnes: Karen Burnes, ABC News, New York.

Protocol for Tracking Discourse, Fear, Crime, and Violence

1. Date of newspaper report or newscast (ABC Evening News): May 3
2. Year: 1992
3. Page and Section/Story Number in Newscast: Story 7 of 9 (7/9)
4. Location of "fear" (or synonyms, e.g., afraid, threatened, dread, alarm, panic, etc.) (We're mainly interested in "fear")
   * a. text—fear
   b. headline/subhead/opening sentence
   c. other
   d. na
5. If Newspaper, Length of Article:
   a. Under 500
   b. 501 to 1,500
   c. 1,501 to 3,000
   d. 3001 plus
   If Newscast, Length of Report:
   a. less than 30 secs
   b. 31 secs to 1 min
   c. 1:01 to 2 mins
   * d. 2:01 to 4 mins (5:44:30 – 5:47:50)
   e. 4:01 mins. or more
6. Subject matter(s) of article/report: (Describe)
   a. contaminants (includes environmental, natural disaster, AIDS, development/developer, housing and building for "others" (e.g., halfway houses, etc.)
   b. economic factors
   * c. other (includes interethnic problems, ethnic groups, ethnic gangs, hate crimes)

The article is related to the riots that occurred after the acquittal of the officers who beat Rodney King. The discussion is about hatred and prejudice and how children can be exposed to those attitudes. It also discusses how to avoid the development of those attitudes in children.

    d. crime and police (includes street crime and criminals, police and official responses)

7. Who or what is "feared"? (Describe)

    a. individuals, includes sociopaths

    b. ethnic gangs

*   c. ethnic groups/immigrants/foreigners

    Discusses prejudice with relation to differences among groups of people and how those differences, and the unknown, can develop into prejudice.

    d. lesbians/gays

    e. developers

    f. facilities for deviants

    g. formal/state agencies of control

    h. natural disaster

    i. other _____

8. Miscellaneous: Include headline, key phrases, relevant 1 to 3 sentences that help identify the report.

"Prejudice is often simply fear. For a child, it is fear of the unknown."

The article discusses how to avoid developing this prejudice in young children.

". . . need exposure to each other so that they can see differences between groups . . . so that they don't stereotype them."

"The kids are the hope of the future, and if they are brought up to think that it's wonderful that there are different races and colors and creeds . . . we'll have a better chance."

9. How is the word *fear* used, as what part of speech (answer for each time the word fear is used):

    a. noun—1

    b. verb

    c. adjective

    d. adverb—1

    e. other

10. Is the story an account where fear provided the reason for doing something?

Fear could resemble an indirect reason for an account as fear could be the reason for prejudice, and prejudice may have contributed to the riots.

If yes, can the account be classified as an excuse or a justification?
Fear is recognized as having the potential to be used as an account because it may be a reason for prejudice. Prejudice and fear may be contributors to the riots in Los Angeles. I am uncertain if people would accept responsibility for the riots and claim that the riots are justified because of the verdict or if they would deny their responsibility for the riots, saying they were out of their control.

A good rule of thumb is that the more familiar you are with the subject matter you are studying, the more precise the categories and codes for each category can be. In these cases, you are able to do more precoding of each category before you actually collect the data. Most coding and commentaries can be agreed on by a team of researchers, although there is never a strong push to use the exact words in describing the report. Generally speaking, the less detailed the protocol, the more likely it is that different investigators will disagree on the relevant categories and codes; conversely, although agreement between investigators is desirable—especially if more than one researcher is collecting data!—it is more important that a single investigator be consistent. The best way to achieve investigator agreement is to work together to not only recode the same documents but also to discuss meanings and interpretations of categories and codes. If disagreements persist, this is usually an indication that there are at least two categories, codes, or dimensions. Make the adjustment and compare interpretations on several more documents.

When data collection is approximately half to two thirds completed, it is a good idea to review your files to overcome any blatant oversights before moving on to full-scale data analysis. The basic approach is to review your files and look for examples of missing or underrepresented categories so that adjustments may be made with the remainder of your sample.

Data Analysis

Step 9. Perform data analysis, including conceptual refinement and data coding. Read notes and data repeatedly and thoroughly.

Step 10. Compare and contrast "extremes" and "key differences" within each category or item. Make textual notes. Write brief summaries or overviews of data for each category (variable).

Step 11. Combine the brief summaries with an example of the typical case as well as the extremes. Illustrate with materials from the protocol(s) for

42

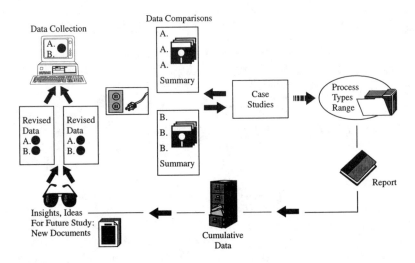

Figure 3.4. Emergent Qualitative Data Analysis

each case. Note surprises and curiosities about these cases and other materials in your data.

Qualitative data analysis may be the most misunderstood aspect of all qualitative research, and particularly document analysis. Steps 9 through 11 illustrated in Figure 1.1 refer to different moments in emergent data analysis. Figure 3.4 illustrates the process of qualitative data analysis and shows how data for common categories are transferred to files (the diskettes) where they are summarized and then compared for key case study analysis. This then permits a tentative finding about the overall process, types, and range of material in each file, which becomes the basis for a report to be cumulatively added to other reports but also as data for further insights, ideas, and future study (Ball & Smith, 1992; Feldman, 1994; Kelle, 1995; Weitzman & Miles, 1995).

The goal of qualitative research is to understand the process and character of social life and to arrive at meaning and process; we seek to understand types, characteristics, and organizational aspects of the documents as social products in their own right, as well as what they claim to represent.

Qualitative data analysis is not about coding and counting, although these activities can be useful in some parts of fulfilling the goals of the quest for meaning and theoretical integration. It is for this reason that I do not try to sell my students on one or more of the coding programs that are available for purchase (Pfaffenberger, 1988; Richards & Richards, 1994; Weitzman & Miles, 1995). The goal is to understand the process, to see the process in the types and meanings of the documents under investigation, and to be able to associate the documents with conceptual and theoretical issues. This occurs as the researcher interacts with the document; only if computer software helps this process should it be used.

My preference is to rely on the more straightforward "search-find-replace" options on most word processing programs. This is especially true for the initial steps in data analysis. Although dedicated coding programs can be helpful, such programs cannot think and they cannot decide the best way to conceptually integrate your materials, yet their operating logic forces researchers to make decisions that may be premature. They cannot deal with meaning, but only common words. Similar words are less important than meaningful patterns that are often apparent in context and relevance to another course of action, which may not even appear in one's notes with the same wording.

In general, data analysis consists of extensive reading, sorting, and searching through your materials; comparing within categories, coding, and adding key words and concepts; and then writing minisummaries of categories. Any technique or software (e.g., word processors and database managers) are fine as long as they permit category-by-category searches, whether done with the entire record or document or individual fields or categories within a record (Huberman & Miles, 1994; Richards & Richards, 1994). For example, with the above news protocol about fear, I initially went through the various categories record by record (i.e., article by article), comparing subject matter. As the approach becomes more familiar, the researcher might try more sophisticated software programs, but the basic steps of searching, finding, sorting, and comparing will remain essentially the same.

The next step in qualitative data analysis is to become very familiar with your data. This occurs through careful and extensive reading (Strauss & Corbin, 1990). Next, while keeping your original file safe, put your various categories (e.g., location, length, source, etc.) in files and read them carefully. Basically you analyze all of the data by comparing specific categories, within and then between categories, and then "typical" cases that have certain thematic and presentational characteristics. Strive for two or three sentences summarizing the range, the extremes, the most typical, and what this suggests at this point in your analysis. After these statements have been made about each category, read the protocols over again, glancing at the original documents. Add any relevant comments to your notes about the document on the hard copy as well as file copy of the document.

Step 12. Integrate the findings with your interpretation and key concepts in another draft.

This step involves summarizing each of the categories in a paragraph, using illustrative materials where appropriate, including descriptions and quotations. Thus, if your protocol consists of 15 items, you would have approximately 15 sections. Next, read this carefully, making notes in the margins. This will be useful in the next step, actually sorting the documents into types with distinctive characteristics. In most cases, the most important will be the variations on themes and focus: If, for example, you have 20 documents analyzed, are there 3 to 5 themes in which most can fit? How do these differ with respect to their source(s), titles, length, and so forth? Which ones do not fit? What is missing or odd about them? What are some surprises from the research? What would you suggest changing for a future study, either with the protocol or the sample?

Protocol-aided document analysis permits the researcher to also keep separate the data and examples from more general conclusions on which they are based. Grounding our assessments of the social world in qualitatively oriented research helps preserve the processual character of social life even as we are able to capture it in analysis. The remaining chapters illustrate how print, electronic documents (mainly television), and even field notes as documents can be accessed, studied, and interpreted using the approach discussed in this chapter.

4. NEWSPAPERS, MAGAZINES, AND ELECTRONIC DOCUMENTS

Qualitative analysis of documents follows the basic steps suggested in Chapter 3, but there are variations due to accessibility of materials, computer facilities available for retrieval, storage and analysis, and the theoretical focus. This chapter presents information about how to collect data for computer analysis from newspapers, magazines, and the electronic forms in which many are now available.

The major news media are central aspects of popular culture, which has pervaded every major social institution, and, indeed, these institutions have adopted much of the logic and format of these media. It is certainly true today that any serious analysis of American life and culture—and increasingly, much of Western culture—must consider media materials.

Many studies, however, have not done this, and one reason is that the difficult access, including finding relevant materials for one's project, have discouraged researchers from using these important documents. In recent years, the major media have become far more accessible. They are becoming more available for analysis as entire collections are being placed on microfiche and more in electronic information bases such as CD-ROMs in public and university libraries.

In addition to regular card catalogues in libraries and microfiche collections, computer on-line terminals in many libraries now permit direct access to newspaper records that are usually within 2 to 3 months of being current. Individual subscribers to some commercial computer on-line services can also gain access to current (within 24 hours) holdings. Indeed, the most sophisticated source to date, LEXIS/NEXIS, contains hundreds of magazines, newspapers, wire services, as well as legal and government documents. The typical format works in the following manner: After locating the newspaper, usually by selecting a computer file, key words may be typed indicating your topic of interest. For example, in one study of the way the word *peace* was used by *Time* magazine over a several-decade period, the word *peace* was entered and articles that focused on it were listed. In some cases, the entire text is available on-line, which can then be read on the computer screen or downloaded onto a diskette. A researcher can then print the article to have a hard copy and also make a backup copy of the article onto another file, on which codes, concepts, and other remarks may be inserted (but marked off by parentheses) to assist in the ongoing coding and analysis. In other cases, however, all that might be available are the title, date, volume, and page numbers. The researcher will have to go the stacks in the library, find the bound volume, and the appropriate article. Ordinarily, it is a good idea to photocopy the article so that it can be taken from the library and marked on for data collection and analysis.

Other resources include documents available through the Internet, the World Wide Web, and other computer services. The capacity to use these sources will be as important as the skill to use a library card catalog in the 1980s. These are like interactive libraries where one can find documents on a wide range of subjects but with the added feature that the author can be contacted directly or other interested parties can discuss topics. They can be downloaded from the main computer to one's personal computer or disk for printing and analysis.

Studying News Magazines

One of the simplest protocols to follow is basically one category, although it also includes the case (or record) number, date, and location. One of my students, Robert Todd, did this for a project I was working on dealing with the different uses and meanings of peace over a six-decade period. The question I wanted to answer was quite abstract, "Has the meaning of peace changed in recent history?" But, following the guidelines discussed in Chapter 3, I could transform my query into a researchable question by asking, "How was *peace* presented in news reports?" And more specifically, "What did *Time* magazine report and how was the term *peace* used in recent history?" After developing a protocol along the lines discussed in Chapter 3, data were collected, checked, and later coded. Because this was an exploratory study, a very simple sampling strategy was used: To attempt to determine how peace was used in different eras, quotes containing the word *peace* were collected for each January issue in 1935, 1945, 1955, 1966, 1976, and 1985 (January 1965 and 1975 were unavailable). Essentially, the use and context of the word *peace* is what was sought in the various issues of *Time*. Two completed protocols are presented below and are followed by several pages from the overall report that illustrates how the data were used. Note the basic descriptive material that can be precoded or coded later.

Date: January 14, 1935
Article #35 A
Location: p. 18; INTERNATIONAL Section
Context: Article described meeting between Mussolini and the French.
French Prime Minister Laval described as a "swarthy, thick-lipped, beady-eyed one-time butcher boy . . ."
Peace: "secret diplomacy as practiced by Benito Mussolini led in Rome last week to an Italo-French 'entente' of the first importance to the *peace* in Europe."
"As Prime Minister Laval prepared to board his train for Paris most observers agreed that he and Benito Mussolini had made each other prime candidates for the 1935 *Nobel Peace Prize*—even if squalling Abyssinia is incidentally butchered."
Laval: " 'we have given rise to great hope, great hope!' continued Orator Laval. 'The world follows our efforts with passionate interest! All who are animated by the *ideal of Peace* today have their eyes turned to Rome! We must not deceive them. *Peace* must be maintained!' "

Date: January 14, 1966
Article #66 D
Location: p. 31 A/B - 32

Context: Article lays out 14 points for successful "peace settlement" in Viet Nam; main point—determining own future "without external interference." North Vietnam must stop aggression on and subversion of South Vietnam. Hanoi denounced the new peace proposals as a trick—wants share of South Vietnam's government.

p. 31 A

"on Veteran's Day the idea of linking halt in bombing with *Peace* talks occurred"

p. 31 B

"While the sounds of war continued in volume in Viet Nam, a remote and quiet Lyndon Johnson sat last week in his oval office waiting for signs of *peace*."

p. 31 B, 32

"Back home, the U.S. *peace offensive* had already struck sparks of domestic debate on the eve of Congress, reconvening this week . . ."

p. 32

GOP Senator Dirksen: " 'We must have capitulation before there is *peace*,' " . . . otherwise, " 'how much negotiation are you going to get?' "

Our discussion about these materials, along with Todd's coding of "accounts," (see below) provided a prelude to further analysis of the data. After a series of section summaries and brief comparisons, as suggested in Chapter 3, a descriptive analysis of the different contexts, uses, and meanings of peace was provided. The following is part of it:

The purpose of this preliminary research is to consider the variations of style, usage, and meaning of the word and concept of *peace*. The fundamental question is whether the word *peace* or the concept of peace has received different treatment over time.

In his work on the vocabularies of motive, C. Wright Mills suggests that talk concerning motive is an important element in everyday interactions and maintains that language serves both social as well as individual functions. Language serves to co-ordinate diverse action and may indicate future action. Thus, imputation and avowal of motives are important. While not focusing solely on the work of Mills and other Symbolic Interactionists, this research will be considered within that framework.

In an era where isolationism was in vogue in the United States, the words attributed to Italy's Mussolini and Germany's Hitler may have helped to diffuse any desire to increase involvement in Europe.

Mussolini, who was meeting with French Prime Minister Laval, was said to have assured Abyssinian Chargés d'affaires Negrades Yassou, "that he wants to maintain *peaceful* relations with Abyssinia and that Italy has not the slightest idea of aggression." Laval stated: "we have given rise to great hope, great hope! The world follows our efforts with passionate interest! All who are animated by the ideal of Peace today have their eyes turned to Rome! We must not deceive them. Peace must be maintained!" Hitler, on the other hand said: "if anyone attacks Germany, he will fall into a hornet's nest because we love freedom as much as we love peace. . . . We give assurances that no pressure, no need, no force, will ever lead us to sacrifice our honor, or the right to equality with other nations!" Here the idea of honor is tied to peace. Thus, peace would be sacrificed by some for honor or other ideals. However, based on the rhetoric of 1935, it appeared that peace for its own sake was considered by many, including the United States, to be a worthwhile goal. Thus, the definition of peace simply as "freedom from war" seems to fit somewhat the peaceful goals of that era.

In 1955, the Cold War seemed to dominate the thinking of the time and the way peace was perceived. The idea of "peace through strength" seemed to dominate the consciousness of the time. In a time where fear of nuclear destruction was a dominant concern, it was suggested that only through shows of strength and resolve could peace be maintained. President Eisenhower stated: "Free nations must 1) maintain and strengthen their alliances against the Communist threat if the insecure peace is to be preserved, 2) negotiate wherever negotiation will advance the cause of a sound peace, and 3) maintain countervailing military power to persuade the Communists of the futility of seeking their end through aggression."

At this point, let us turn from the text to another analytical focus, the concept of "accounts," referred to in Chapter 3, that has proven very useful in news and document analysis (Scott & Lyman, 1968). The data about peace were coded as excuse or justification. The following is how Todd drew on the material he coded as either an excuse or justification:

In an attempt to explain militaristic action, leaders utilize either excuses and/or justifications, two techniques suggested by Scott and Lyman. An actor utilizing an excuse admits that a behavior is problematic but attempts to explain that behavior away by denying responsibility. For example, perhaps in part to excuse chilly relations with Russia, Dulles reminded the world that "after nine years of delay and diatribe, the Soviet Union still refused to sign a peace treaty ending the occupation of Austria." The use of justifications, on the other hand, is a common means by which an actor accepts responsibility by explaining that a seemingly problematic action is not really wrong. From this research, justifications seem to be the preferred mode of explaining behavior, because government actors apparently wish to avoid any suggestion of impropriety. Justification occurred when, for example, the United States government justified their cold war attitudes by seeming to suggest that any negative action taken was only a logical, necessary reactions to the Communist menace.

Justifications were also in vogue in 1966, when the Viet Nam conflict dominated the pages of *Time,* and President Johnson's "peace offensive" was the story of the hour. The combination of the word "peace" with the aggressive, militaristic term "offensive," may foretell the problems facing Johnson in bringing an end to the war. The term peace offensive assumes the tone of an oxymoron when taken in conjunction with the quote that Johnson "made it clear that the peace offensive will continue, while leaving no doubt that the U.S. will stay in Viet Nam." While Johnson sought peace, he could justify the killing and U.S. intervention in Viet Nam based on the duty of the U.S. to protect the South Vietnamese. Again strength was more important than a tarnished peace. Senate Minority Leader Dirksen said: "Let the peace efforts continue, let the military effort continue. It demonstrates our determination to keep our word." The belief that peace is probably unattainable without war is also a viewpoint expressed more frequently in later decades of the study. "Johnson himself acknowledged implicitly that his peace overtures had come to naught for now. 'I think every schoolboy knows,' said the President . . . 'that peace is not unilateral—it takes more than one to sign an agreement. What is holding back peace is the mistaken view on the part of the aggressors that we are going to give up our principles, that we may yield to pressure, abandon our allies, or finally get tired and get out.' "

In his concluding section, Todd reviews the protocol data and reflects on the changing notions of peace in the context of changed history, including the changing focus of the magazine:

Moreover, despite exclusive usage of *Time* magazine, changes in the style, usage, and meanings of peace are reflected in part by changes in format and style of the news magazine. In 1935, for example, the style of *Time* was basically to report the news as quickly as possible. While choice of words was clearly slanted in favor of the United States' interest, articles read primarily as a rambling travelogue. In 1945, *Time* reported thoroughly the "facts" of the war, but again with a slant toward patriotism and "keeping the home fires burning." Again in 1955, the magazine seemed gripped in the same cold war fever that was apparent in the words and actions of many world leaders. By 1966, some changes occurred in context and format with an increasing emphasis placed on analysis. With the Viet Nam conflict coming into American homes on the evening news, explaining events thoroughly, not merely reporting, became important. In 1976 and 1985 in-depth analysis and explaining the effect of the news on the average person became vitally important. In an increasingly complex but shrinking world people had ready access to the "facts" but needed a friend to interpret the meaning of those "facts." News magazines such as *Time* have succeeded in part by adjusting to the times, just as politicians and world leaders have succeeded by modifying motives to what they believe people want or need to hear.

Studying Newspapers

A little more elaborate protocol was used in another student's study of newspaper reports of Arizona's first execution of a prisoner in some 30 years. Crime news, of course, is the staple of local news organizations; indeed, a survey of virtually any local newspaper or TV newscast will show that there is little local news except for crime (for a discussion of the implications of this for social control, see Ericson, Baranek, & Chan, 1989.) As with most protocols, this one provided data that could be used for quantitative and qualitative analysis. Robin Ann Rau (1993) essentially followed the guidelines suggested in Chapter 3, reading studies of the news process, and particularly the role of news sources, news frames, and news formats in packaging reports. She also read a number of newspaper articles about several issues surrounding the execution of Don Harding, a convicted murderer. For theoretical reasons, she sought to examine which sources were associated with certain themes and frames. The next step was to show how these themes and frames were associated with news sources and also to note how they changed or "moved" over a period of time. And these may fluctuate over the course of an entire event report. It is this time frame or scheduling of themes, frames, or sources that needs to be more incorporated into your sections, especially the ones on themes and frames.

Rau obtained the news articles from an information base of the *Arizona Republic*. In most computerized information bases, the name of a specific publication—for example, *Time* or *Arizona Republic*—is referred to as a "file" that is usually part of larger "library" or collection of magazines or newspapers. The unit of analysis was the newspaper article, but the protocol was applied to each instance involving an identifiable source because Rau was particularly interested in the relationship between sources, themes, and frames. Thus, an article may have more than one source and therefore several data points or applications of the protocol.

On reading several articles, it became apparent that not all articles about the Don Harding case had his name in them, but rather, they may just refer to the "history of executions" in Arizona and the like. This is important because if Rau had not been familiar with the articles, the language, various interest groups, and competing sources and "claims-makers" involved, a number of important articles would have been omitted had her main search term been *Don Harding*. Part of the task of informed document analysis is to be familiar enough with the publication(s) providing research materials and the major terms and concepts used so that few articles will be missed. For example, she found that *gas chamber*—Arizona's choice of "justice"— identified several articles about the Harding case. Rau selected 32 articles, copied them to a disk, and printed them. With repeated reading she became familiar with the reports, revised the protocol to suit her emerging focus, further "tested it" against other articles, and finally settled on the following:

Date of Article
Location of Article
Size of Article
Source/Agency Cited
Language Describing Execution
Definition of the Problem
Message/Emphasis Source Is Sending
Viewpoint in Maintaining Order
Themes: Policy, Mitigating, Pain
Frames: Legal, Justice, Personal, Public
Angles: Appellate, Bureaucratic, Legislative, Operations
Moral, Victim, Equity
Brief Description:
THIS WOULD REPEAT FOR EACH "VIEWPOINT IN MAINTAINING
ORDER" IN THE ARTICLE

Included in these articles were 89 sources providing information, claims, and points of view. The latter were coded as various themes and frames. Four frames emerged in Rau's study: legal, justice, personal, and public. Extensive descriptive materials were recorded in the protocol for each report. This creative approach to a series of news reports was crucial in documenting how the "story" changed over a period of time; it had a "career" that was tied to certain sources' perspectives and attendant themes and frames. Three themes emerged: policy, mitigating circumstances, and pain. Studying the documents indicated that certain perspectives and general points of view were evident that basically cast any discussion along certain lines and not others.

Rau's review, coding, and analysis led her to employ several analytical aids that were also useful in data presentation. Although she made extensive use of headlines and quotations from sources in the various sections of her paper, she was also able to use frequencies of reports in several ways. One was to construct a bar graph showing the number of reports associated with certain sources. She was interested in the career and development of the "Harding story" over time, so she also constructed a chart indicating which themes (and how many) appeared in news reports at different periods of the reportage. The ensuing reports were less about Harding than about the source's vantage point. For example, in her conclusion she wrote, "Don Harding was depicted as a man suffering from a 'brain impairment' who lacked the 'ability to conform his conduct to the requirements of the law'. . . and on the other hand was a 'cold blooded, cruel, vicious killer.' With each source's report, the stage for the Don Harding execution was tailored from the source's use of frames and angles."

Exploring Electronic Documents

Another mode of analysis is possible, however, when numerous documents are available to search and analyze. The new form of documents permits the simultaneous development and exploration of documents. One no longer has to restrict searches to one document at a time, but can find cases and collect information from thousands of documents. One example is a project I call "gonzo justice," the use of extraordinary means to demonstrate social control and moral compliance, often through rule enforcement and punishment designed to stigmatize publicly (e.g., the mass media), and to demonstrate the moral resolve of those mandating the punishment. When judges impose "outrageous" sentences designed to stigmatize the accused and also call media attention to the judge's act, this is gonzo justice (Altheide, 1992). Consider a few examples. The following one is from Pennsylvania:

A 311 1/2 pound man who hasn't made child-support payments for more than a year because he's too overweight to work is under court order to lose 50 pounds or go to jail. . . . "I call it my 'Oprah Winfrey sentence,' " [Judge] Lavelle said. . . . It's designed to make him lose weight for the benefit of his children, while Oprah (a talk-show host) lost weight for the benefit of her job and future security. (Thorpe, 1984)

And another is from Tennessee:

A judge ordered Henry Lee McDonald to put a sign in his front-yard for 30 days declaring in 4-inch letters that he 'is a thief.' U.S. District Court Judge L. Clure Morton instructed McDonald to erect the sign Tuesday as part of his three-year probation for receiving and concealing a stolen car. . . . The sign must be painted black and have 4-inch white capital letters that read: 'Henry Lee McDonald bought a stolen car. He is a thief.' ("Thief," 1989)

News reports similar to this can simultaneously contribute to our understanding of social control, the role of the mass media in the drama of everyday life, and how to evaluate documents and sources of information. But they are rare and, until recently, have been difficult to find and study systematically. Indeed, the examples presented above were found quite accidentally over a several-year period. The latter consideration entails, first, developing a general approach to relevant information as documents and second, placing these documents in a context of meaning to interpret them. These points have emerged from some recent efforts to develop a conceptual framework for the qualitative study of documents. I call this phase *the double loop of analysis.*

The Double Loop of Analysis

What is significant about these reports is that the word *gonzo* does not appear in any of them because it is a construct borrowed and elaborated on from the journalist Hunter Thompson and conversations with my colleague, John Johnson. I collected numerous examples of gonzo over a 5-year period before I knew what to call it.

We can develop some common terms and phrases of gonzo from the examples to date and then conduct searches of literally thousands of news documents. The emergent process is multifaceted and therefore difficult to set forth linearly, step by step. The gonzo justice study illustrates, however, how it partially occurred through a double loop of analysis. The first loop involved exploring some articles that seemed odd in the way they reflected moral and personal statements by judges. Some of the articles are presented above. But it also involved describing and comparing several of these after they had been placed in my word-processing database. Looking for commonalities led to the definition of gonzo justice, which in turn sensitized me (looped back) to some general features of other articles, which I was able to discuss with colleagues, who then brought me several other examples. This is when the second loop began to operate. The second loop involved further clarification of the common principles and then applying them to the search criteria of several information bases. In this phase, the information base and its logic led me to examine the data to look for some initial commonalities that would facilitate additional data collection and specification. Among the referents were formal agents of authority (e.g., police and/or judges) making very explicit moral assessments of the accused and the act, stating their personal disdain, and adding that their sentence was intended to do more than fulfill the letter of the law; it was to *instruct, teach,* and set an *example.* Although all of these words were not contained in each case, the clarification led to the focus on the examples available.

Some terms common to many of the examples of gonzo justice are *judge, orders, lesson, and probation and parole.* These common terms were used in other searches in NEXIS. Preliminary excursions in the massive NEXIS news information base for the last 2 years of publications using these terms in various combinations have been encouraging. For example, this search strategy found the following reports that were read, preliminarily assessed, and then downloaded onto a disk for later investigation and analysis. The first is from the *Los Angeles Times* ("Scales of Justice," 1994).

> A judge who wanted to punish a piano teacher for molesting two girls decided to do so in a way that would affect the defendant most. He has ordered George Marrs to stay away from the keyboard for 20 years.
>
> That condition was part of the probation that state District Judge Ted Poe imposed as part of a plea arrangement that kept the musician out of jail for fondling 9- and 10-year-old girls.
>
> "I took away the most important thing to him," Poe said. "I don't know anything about pianos. I don't like pianos. But to him, that piano was the most important thing in his life."
>
> Marrs, 66, was accused of molesting the girls at his home during piano lessons. Both girls eventually told their parents, who notified authorities.
>
> Marrs, who pleaded no contest Wednesday, also must donate his $12,000 piano to a children's home and post a sign on the front door of his home that warns anyone under 18 to stay away.
>
> Poe conceded Thursday that Marrs might get away with playing a piano somewhere other than his home. The intent of the unusual probationary clause, however, is to keep him from ever owning a piano again, thus stopping him from giving lessons, the judge said.
>
> A probation officer, in periodic visits to Marrs' home, will check for pianos, Poe said.
>
> "It is certainly appropriate because there is a nexus between the crime and the punishment," Poe said. "I see no legal problems at all. It's a means of keeping him from possessing a piano."

What is very interesting about this document is that it is the first instance I have found in which a judge clearly admits that he is trying to punish an individual even if it is not directly related to the offense under review. And this is in addition to the order to place a sign on his door. This is just one case, of course, but it suggests, using some of the coding categories from prior work on gonzo justice, that this approach to punishment has been taken another step. Compare this more pronounced interest in punishment with another case found with this new search procedure.

The *St. Petersburg Times* published the following gonzo report ("Judge Orders Lesson," 1990):

Three young men who spray-painted "Hitler is Back" on a high school wall were ordered to learn about the Holocaust by a judge who noted that Tuesday was the anniversary of V-E Day. Westchester County Judge John Carey ordered the three to return to court June 25 for a quiz on their history lesson. He also sentenced them to weekly tours at a hospice for the terminally ill. "I don't wish to punish any of you," Carey said. "I wish to educate you." Michael Morano, 18, Michael Venturino, 18, and Neville Anthony Mason, 20, all of Ardsley, were charged with spray-painting swastikas and writing the words "Jew" and "Hitler is Back" on the walls of Ardsley High School last June. If the three fail to pass the quiz, the judge could give them the exam again later or consider it a violation of probation and resentence them.

Noteworthy for my ongoing study of "gonzo justice" is that in the earlier 1990 case the judge clearly stated that he was not trying to "punish" but was trying to "educate" the youths. The 1994 case, as noted, was trying to "punish" the accused. Although additional study will be undertaken to sort out the social significance of this kind of justice, the important point for document analysis is that there is now a way to mine the thousands of news publications from around the world to monitor the character, frequency, and use of gonzo justice. This process was very useful in developing a research strategy to track discourse and seek general patterns beyond gonzo (see Chapter 6).

Print Media Photos

Most document analyses are oriented to written text, although more of our public information and popular culture involves visuals. The most significant type of visuals in our postmodern world is electronic, especially television, and the next chapter focuses mainly on analysis of TV documents. The electronic media have played such a large role in social life that they have influenced the formats of numerous print media as well. More visuals are included in newspapers than previously, and this is true of magazines as well. Because visuals contribute to the meaning and "look" of information and content, it is important to have a strategy for analyzing photos and other visuals qualitatively.

There are various approaches to this task, including some very creative work in semiotics, that look for underlying messages or cultural "signs" in visuals that can then be related to "sign systems" (Ball & Smith, 1992; Berger, 1981, 1982). Even this more abstract approach, however, must begin with a careful description of the visual and its immediate contribution to the meaning and format of the article with which it is associated. Because of space limitations, then, I will make a few suggestions for studying news photos that are related to a news text. Although many of these points will also be helpful in analyzing advertisements and cartoons, they are not the primary focus.

As with all news reports, news photos appearing in newspapers are a result of a long process or career. In qualitative analysis terms, one can say these photos are "reflexive" of the process that has produced them, including numerous decisions by photographers, photo editors, caption writers, and city editors. In general, visuals in newspapers are not the primary defining feature of what is selected as newsworthy: Photos are taken to document, illustrate, and support a news story.

The challenge for qualitative document analysis of news photos is to capture the meaning, essence, organization, and relevance of the photo for the news story. To put it differently, the researcher would prefer to be able to obtain data from a report to document that this is "what the story looked like." I have found that news photos are best approached with only a few categories to be coded after the data collection is well underway or completed. In the context of a newspaper report, for example, the following codes were used in constructing a protocol for a study of news photos pertinent to the Persian Gulf War in the *Arizona Republic:* source, date, location, size of photo, photo caption, photo source, whether a specific article was related to the photo, the source of the article (e.g., the Associated Press), and a detailed description of the photo. Illustrative completed protocols are provided in the following:

Photo #41
Source: *The Arizona Republic*
Date: Tuesday, January 15, 1991
Location: Page A1, upper left-hand corner
Size: 5" × 5"
Photo Caption: "Gen. Norman Schwarzkopf reassures a soldier in Saudi Arabia of U.S. support for the military effort. On Monday, demonstrators in the United States turned out for rallies to support soldiers."
Photo Source: David C. Turnley/*Detroit Free Press*
Article Related to Photo: Not specifically related to any photo.
Article Source: None
Photo Description: General Schwarzkopf is standing with another soldier next to what appears to be a military vehicle. The General has one hand resting on the soldier's left shoulder. Indicated by the smile on General Schwarzkopf's face, it seems that the conversation between the two soldiers is quite light.

Photo #42
Source: *The Arizona Republic*
Date: Tuesday, January 15, 1991
Location: Page A1, upper right-hand corner
Size: 5" × 5"
Photo Caption: "Chicago demonstrators burn a flag in protest of a possible war. With the deadline for a peaceful solution to the Gulf crisis looming, antiwar rallies took on a desperate edge Monday."
Photo Source: John Zich/The Associated Press
Article Related to Photo: "Global throngs pray for peace, assail likely war"
Article Source: The Associated Press
Photo Description: Simply a picture of a crowed of demonstrators burning an American flag. Behind the burning flag are other demonstrators holding signs and banners.

These categories permit the flexibility to do a reflective coding as the data analysis proceeds. What I mean by the term is that the researcher moves back and forth between protocols as data in one category suggest some similarity or difference from one just read a moment or two ago. Through the first two or three readings of the data, notes should be jotted down about the data and particularly when codes emerge. For example, in going over the data for more than 100 photos in the *Arizona Republic,* certain codes emerged that were consistent with other studies as well as theoretical work in the media. These codes suggested corollaries or their opposites, in a relational sense. For example, just looking at the descriptive data, as well as the photos themselves and the articles to which they were attached (literally, stapled), the photo descriptions suggested the following range of topics: (a) equipment and weapons; (b) equipment and weapons with troops; (c) humans and troops only; (d) U.S. and "our side" soldiers; (e) enemy, including soldiers; (f) enemy civilians; and (g) antiwar demonstrators. An additional category is the "scene or setting" in which the photo was taken and what it is presumed to represent (e.g., battle scene). Several could be coded with the relevant category for each of the protocols. The same procedure was done with sources (e.g., mainly government/military, wire service, other U.S./coalition media, television, Iraqi, other). And the emphasis in captions was also "reduced" to some common codes, including themes. The codes were then applied to the protocols as suggested in the revised protocols.

Photo #41
Source: *The Arizona Republic*
Date: *PRE-WAR* Tuesday, January 15, 1991

Location: *DOMINANT* Page A1, upper left-hand corner
Size: 5" × 5"
Photo Caption: *EMPHASIS MILITARY UNITY, SUPPORT* "Gen. Norman Schwarzkopf reassures a soldier in Saudi Arabia of U.S. support for the military effort. On Monday, demonstrators in the United States turned out for rallies to support soldiers."
Photo Source: *OTHER U.S.* David C. Turnley/*Detroit Free Press*
Article Related to Photo: Not specifically related to any photo.
Article Source: None
Photo Description:*MILITARY/EQUIPMENT* General Schwarzkopf is standing with another soldier next to what appears to be a military vehicle. The General has one hand resting on the soldier's left shoulder. Indicated by the smile on General Schwarzkopf's face, it seems that the conversation between the two soldiers is quite light.
SCENE Everyday interaction, informal.

Photo #42
Source: *The Arizona Republic*
Date: *PRE-WAR* Tuesday, January 15, 1991
Location: *DOMINANT* Page A1, upper right-hand corner
Size: 5" × 5"
Photo Caption: *EMPHASIS ISSUE, CIVILIANS* "Chicago demonstrators burn a flag in protest of a possible war. With the deadline for a peaceful solution to the Gulf crisis looming, antiwar rallies took on a desperate edge Monday."
Photo Source: *WIRE SERVICE* John Zich/The Associated Press
Article Related to Photo: "Global throngs pray for peace, assail likely war"
Article Source: The Associated Press
Photo Description: *U.S. CIVILIANS, ANTIWAR* Simply a picture of a crowd of demonstrators burning an American flag. Behind the burning flag are other demonstrators holding signs and banners.
SCENE Demonstration, public symbolic act, intended to be viewed by an audience.

When this initial coding is completed, the different categories and codes can be grouped for additional comparisons as suggested in Chapter 3. Frequencies of codes might be noted as part of the initial brief summaries to be written for each category and code, including the most common photo descriptions, the least common, and with good illustrative descriptions. In general, preliminary analyses of these data (including the cases not reported here!) indicate that the majority of the photos pertained to equipment and weaponry, with occasional connections of troops-to-weapon systems. There were very few photos and scenes of individuals, particularly those of enemy civilians, although there were a number of Saddam Hussein, the arch-villain in Western mass media propaganda. For example, in 102 photos published in the *Arizona Republic* between January 15 and January 31, 1991, there were at least 8 photos of Saddam Hussein, although his name was used at least 3 times as often in photo captions. Incidentally, there were 4 photos of troops "personalizing" messages "for Saddam Hussein" on bombs and missiles that were aimed at Iraq! This brief illustration of qualitative photo analysis suggests some of the rich materials that can be obtained through a minimaᶦ application of categories, study, emergent codes, systematic comparisons, and analysis.

5. ELECTRONIC REALITY

Studying television has never been more important or challenging than it is today. Whether one agrees with a contention I and others have tried to establish that the mass media and especially television are the most important social institutions in the Western world, there remain ample reasons to study some aspect of TV and other electronic media. Whether it is referred to as virtual reality, hyperreality, or even cyberspace, electronic experience has become part of our daily lives. The key difference from other media in other ages is the power of the visual and the capacity to interact with it, even change the experience one is watching by treating it as a game. We know about events as news reports, fashioned by claims-makers, and media formats featuring entertainment. These mediated reports become reality and shared frames of reference for public issues. Chayako (1993) stated it as the following:

> To use a continuum [subjective-objective reality] rather than a rigid frame to conceptualize the organization of experience and reality itself is one way to become more sensitive to the ways in which actors respond to the new complexities of the real, near-real, hyperreal, unreal, pretend, and so forth. . . . In the case of virtual reality, we must investigate the processes by which actors frame and interpret the experience before we make inferences about its nature and impact upon society. (pp. 178-179)

It is no wonder that many students of social life argue that virtual reality has become the dominant reality for many people! Whether the context be "serious," "work," or "play and entertainment," the popular media are increasingly part of our everyday lives, which are more and more electronically oriented.

The complete study of electronic media, their logic and formats, and their social consequences for social order has occupied the attention of many for several decades. Some of those efforts extend beyond the media materials and content to the actual integration of media perspectives into our major social institutions and everyday activities. Many of these materials are cited in the References and in the Appendix and will not be covered here, but the study of documents and content is a significant feature of this work.

TV Materials

Television as well as movies are now easier to obtain and analyze because of changes in information technology (e.g., CD-ROMs, on-line information bases, and videotape). There are several ways to obtain television news materials. One way that is quite recent and affordable is to videotape newscasts about the topic one is interested in or the entire newscast(s) for a week or more.

At least two other sources of information about TV news are useful for researchers: the Vanderbilt University Television News Index and Archive, and various CD-ROM information bases, but especially on-line systems such as LEXIS/NEXIS. Begun in 1968, the Vanderbilt archive is an expanding collection of network nightly newscasts, plus other network "special reports" and some "event reports," (e.g., the Persian Gulf War). Newscasts are taped, and then each network's newscast is annotated one news report at a time, complete with beginning and ending time, the correspondent, and whether visuals were used. Commercials are also noted. The text of these reports is published monthly, complete with an index.

There is also an annual index so, for example, one could find all reports dealing with "drugs" in a given month or year. Many university libraries subscribe to this index, although it is also available on the Internet through a Vanderbilt University gopher server (TV news.Vanderbilt.edu). Several menus appear for your selection. The best approach is to log on and explore what is available. For a modest fee, researchers can order copies of tapes of newscasts or for a few dollars more, order compiled newscasts of certain events (e.g., only those portions of newscasts pertaining to the Iranian hostage crisis). Written transcripts and audio recordings can also be ordered using forms in the back of each issue of the index. For example, during my work on the Iranian hostage crisis, I ordered specific excerpts selected as part of my sample. I then analyzed the tapes with a protocol that emerged while viewing a portion of this material.

I referred to the Internet gopher server at Vanderbilt University. This has an advantage over the printed indexes because of the "search" option. The researcher can either look at the transcripts (no visuals are available at this writing) by date, month, and year, or he or she can use the search option to find reports on particular topics (e.g., terrorism, Iranian hostages, the drug war, etc.). A screen will appear—there may be several depending on the topic—with the documents. One can then simply select a document, read it, obtain data from it, then download it to a disk to be placed in a computer file for future reading, coding, and analyzing. There are several ways to save these documents depending on the computer system you are using. I recommend saving the data source—for example, the article or the annotated news report—so that it can be carefully examined and then reexamined as ideas and codes emerge. Although this can be done from most university libraries, it can also be accessed from a home or office computer with a modem and appropriate software. This availability is a boon for a serious researcher of TV news or for an undergraduate student who does not like to spend a lot of time getting research materials. Of course, make copies of everything!

Other useful sources of TV transcripts are several commercial, yet often publicly available, information bases, such as LEXIS/NEXIS. Unlike CD-ROM systems that consist of a computer disk with information (usually available in a library) that is updated on a monthly or bi-monthly basis, LEXIS/NEXIS is on-line from a main computer and must be subscribed to by users who are issued a password. Its format is similar to CD-ROM systems that are available in many libraries. Initially costing several hundred dollars per hour, LEXIS/NEXIS is now made available at no charge to the user by a number of universities and public libraries.

Initially established to provide legal materials (case law, Supreme Court cases, federal and state laws, etc.) for attorneys, LEXIS was greatly expanded—adding NEXIS—into news and public information and now includes the major wire services in the world, major newspapers (and numerous minor newspapers), as well as several files for TV news, including *ABC News* since 1988. These files consist of transcripts only (at this point), but they have the advantage over other sources of transcripts because they can be so easily found by a specific topic, retrieved, and downloaded and saved to a disk for later analysis.

It is the way LEXIS/NEXIS is organized and the search options that make it a revolutionary research and teaching resource. LEXIS/NEXIS is organized in "libraries"—combinations of many files—and "files," usually one specific news source (e.g., ABCNEW, which is its file name), but sometimes several will be combined into super files (e.g., MAJPAP, which includes about 30 of the world's major newspapers!). After one signs on and is accepted, the software requests that you select a library and then a file. You are then requested to type in your search request. The request must follow certain rules of syntax that are not too complicated. And files and libraries can be changed quite easily with the use of function keys or "dot commands"—for example, .cf means change file.

The search commands follow Boolean logic, meaning that there are certain logical dos and don'ts. One can search for specific terms, and the search can be restricted and expanded in certain ways. For example, a particular term can be searched (e.g., *homicide*) or a word can be looked for within (w/n) words of another (e.g., homicide w/n women). You can specify if you want to search the entire document or if you just want the term by "segments" (e.g., if it appears in a headline, a certain section of the newspaper, etc.; the relevance of this for TV news will be noted in the following). If, for example, I wanted to check NEXIS for news reports relevant to gonzo justice, I could simply type *gonzo justice*. This would not find any cases (I checked!) because, in that this is a sociological construct, there is not an instance in which those two terms appear together in the file and library where I was looking. As noted in the discussion about gonzo in Chapter 4, however, if I checked for the phrase "judge orders," I would find over 10,000 cases! Examining just a few of these cases made it clear that this was too unwieldy and that virtually none of these cases was on my conceptual track. As noted in Chapter 4's discussion of the double loop, I then returned to my examples of gonzo justice and examined the language more carefully. With a bit more conceptual clarity, I then adjusted my search terms and added another word, *lesson.* My search then became "judge orders lesson," and this turned up a half dozen reports. On examination, it was clear that two in particular were relevant to my use of gonzo, and those were the ones noted in Chapter 4.

Once a "hit" occurs, LEXIS/NEXIS gives you several options for how to display the hits. One is *cite,* which permits you to see the citation and headline; another is *kwic,* or key word in context, which provides the search term(s) and the sentence or two in which it occurs; another is *full,* which provides the complete text of the report on screen, although you will probably have to pass through several screens. The usual progression is to put in a search term(s), check the cite, select a few for kwic, and from these select a few for full. If these are what you are seeking, an individual page or two can be printed one screen at a time or the entire document can be downloaded to a disk to place in a file for later analysis.

Downloading and Storing Files

It has been mentioned that media materials can be downloaded to a disk and then used in later analyses. Obviously, this mainly applies to electronic files, or news materials that have been obtained from a computer information base of some kind. If one is using paper documents in the study, there will still be files to maintain, of course, but not electronic ones, unless the materials are transferred to a computer format. This is not, however, out of the question. One way to do this is to simply copy and transcribe by typing the printed text into a computer file. If there is a lot of text, however, this can take some time. Another way to do this is to use a scanner, which is similar to a photocopy machine that "reads" your paper text and transfers it to a computer disk. Current technology is still quite crude and the copying includes many mistakes that mess up the text, especially if the original is newsprint, which tends to produce a messy or "noisy" computer image.

Putting files in an information base, whether in a computer or in a filing cabinet, is an essential step to collecting data from them. Computer files have an advantage, however, because they can be used to assist in the emerging process of research focus, protocol construction, and, ultimately, data collection and coding. I prefer to make copies of my files for data collection purposes. If a document is in a computer file (although you might actually read a hard copy), it is easier to make notes in the computer text, even conceptual queries to think about later. When that file is saved, so are the notes, and this is very handy so that they can be searched in the future, perhaps after a few more documents have been read and other ideas emerge.

Computer files of documents also permit another kind of analysis that was extremely cumbersome—and, therefore, seldom done—when using paper documents and filing cabinets. This is "multiple document analysis" or the simultaneous study of all your documents. This is usually done by searching all your documents using a few search phrases, which might be simple words, linguistic statements, codes, or concepts that have been entered. Occasion may arise to do this after you have reviewed several documents as noted above or even well into your data collection and analysis of protocols. A term or phrase may come to mind as relevant, and a search can easily be done. Again, it is the emergent aspect of data analysis that is important.

It is an advantage to have the capacity to search documents, especially those that have been "embroidered" and "analytically enhanced" with researcher's notes. Until recently, the main way this could be done was to have each document in its own file, although expanded computer memory and storage might permit the inclusion of several documents in one file. (Of course, the first problem was to remember the name of the file!) The files would then be searched one at a time. If all documents can be placed in one file, that makes them even more accessible. One might have a number of large files, however, and, moreover, several might not even be part of the immediate project, even though they could have relevance for the subject under investigation. The question is how to get access to all of the files at once, especially those that may be more relevant than the researcher thinks—even if you do not recall the name of the file or the subdirectory where you stored it. Recent software development in word processors and database managers has helped solve this problem of how to manage and study multiple files at once. The solution involves indexing or having your computer "read and mark" your files so that a few simple commands can basically treat everything stored in files as "one big file"! Once the files are found, the relevant term can be located, marked or bracketed, and then saved to another file. The system I have found useful is called Zyindex, and there are different versions of it. There are others as well, but this capacity to search all relevant documents is a terrific way to scan and review materials that often become a source of creative insights, codes, and especially examples. Old data look different and usually much richer when viewed through the insights and experience from additional research and emergence. This is particularly applicable to television and other electronic documents.

Analysis of TV Reports

One of the most ambitious tasks is to study the text of TV news, but particularly the visuals. More researchers now agree that the visual formats of TV news are very important in the selection and production of news, although some continue to argue that the "TV talk" or "script" is most important (Fields, 1988). It has taken researchers a long time to get a conceptual handle on visuals, but the guidelines I offer, along with helpful insights from students of semiotics and other dimensions of cultural analysis, now permit systematic study of TV messages as texts. As previous work suggests, the TV format, or guidelines for defining, selecting, and presenting material, is geared to an entertainment approach that includes visual information featuring motion, drama, and action that is also very compatible with conflict and violence.

Finding a way to study visuals is challenging. Recent work on TV news coverage of the Persian Gulf War benefited from drawing on interviews with journalists in four countries, as well as analysis of significant portions of their news reports, including visuals (Altheide, 1995). A comparative study of the news coverage of several wars was undertaken. Indeed, theoretical and empirical work by several colleagues in the United States and Europe suggested that war coverage had differed (Morrison, 1992). The Vanderbilt archive was the major source for obtaining news reports about the U.S. invasions of Grenada and Panama, and the fall of the Berlin Wall and the political changes in Eastern Europe (Czechoslovakia and Rumania). Other materials were obtained from researchers who studied the news coverage of the British invasion of the Falkland Islands (Mercer, 1987).

Reviewing several of the news tapes and reflecting on previous studies indicated that the optimum unit of analysis for data collection would be the individual news story and not the entire newscast. Earlier qualitative analysis of TV news, however, indicated that specific news stories consist of specific parts, called *segments* or *information units*. These are defined as situations in which some "actor" (a person, agent, state) takes "some action" toward an "object" (person, place, or thing). One can see the importance of having videotape to play, replay, and replay again, often 8 to 10 times. Below is the protocol for 2 of the 11 segments of the *ABC Evening News* on December 20, 1989. The first segment lasted 2.5 minutes and the second, 3 minutes. They illustrate how the data look, especially the visual information. Some of this material forms the basis for the brief comparisons about the news coverage of various conflicts. Analyses of news coverage of the Falklands War (Glasgow Media Group, 1985; Mercer, Mungham, & Williams, 1987; Morrison & Tumber, 1988) and Grenada (Mungham, 1987) as well as other conflicts suggest that the political context in which journalists find themselves often informs the coverage (Bennett & Paletz, 1994).

1. Event: Panama; Rumania; Soviet Union (Lithuania)
2. Date: 12-20-89 (Entire Broadcast)
3. VCR counter: 0104-1580
4. Network: ABC
5. Program: *World News Tonight,* Peter Jennings
6. General Report Topic: Occupation of Panama, Noriega at large, casualties; Rumania: demonstrations with violence and destruction; Lithuania: first formal split in Communist party.
7. Eleven Segments—Total time 21:40 (ENTIRE BROADCAST) (plus 10 sec. ad for Nightline)

Segment 1—2:30 (30:00-32:30)

    a. topic: casualties ("provisional figures"); access of news people; Noriega at large; a woman was killed—emphasis

        U.S. claims to have made significant progress, but occupation not complete, Panama City not yet under U.S. control

        sharp, vigorous exchanges between U.S. and Panamanian forces

        cameramen kept as far as possible from action

    b. themes: good v. evil

    c. metaphors: "at large" (Noriega); "sharp exchanges"; "progress in Panama;" helicopters "ferried men in"—the key here (Noriega at large)—caught in crossfire—Noriega's BEHAVIOR INSPIRED INVASION—reel in shock—unfolded throughout the day

    d. sources: U.S. govt; the "military"; "National Security Council"

    e. Types of visuals: actuals: helicopter (shown before anchor comes on for first time—lead-in from insert from local news); black smoke; planes; soldiers on ground; anchor at desk in newsroom then solo; buildings surrounded by black smoke; men running with guns; soldiers in field; soldiers aiming guns; helicopters in sky; airplanes; civilians watching helicopters; civilians walking; bus

Segment 2—3:00 (32:30-35:30)

    a. topic: "Noriega dangerous"—initial attack —thought would capture/kill Noriega quickly

        other forces secure airport, paratroopers dropped in darkness (in Oct. foiled attempt to overthrow Noriega)

        massive force in predawn attack, light tanks over several hours

        ELECTRICAL PLANT, DAM SECURED

        NUMBER OF TROOPS

        Am. Troops thought could capture Noriega quickly

        civilians afraid of armed thugs

        Operation Just Cause

    b. themes: strong v. weak—"massive attack"; good v. evil (Operation Just Cause)

    c. metaphors: "Operation unfolded"—"H Hour"—"from one end of Panama to the other"—"decapitated HIM from dictatorship"—Noriega is "fugitive"

    d. sources: "no one in Panama"—"Pentagon officials"

    e. visuals: anchor with graphic labeled "THE ATTACK" with flag/arrow pointing to Central America

        actuals: nighttime, smoke over buildings; bombs exploding

        graphics: U.S. Southern Command and map with arrow moving to show movement of troops; marine insignia; map of airport—Central America

file film: men walking with hands above heads in front of building
(Panamanian troops foiled attempt to overthrow Noriega in Oct.)
graphic: maps U.S. and Panamanian flags—show where prisons;
actuals: soldiers walking; helicopters; black smoke overbuildings; Sec.
Defense and Chr. Joint Chiefs at Press Report
civilian men running; reporter

In addition to the unit of analysis, event, date, tape counter (for reviewing the report later), network, and program (each of which can be precoded), our emergent studies indicated that it is also important to obtain information about the general report and its topic. Then specific subprotocols would be completed for each of the segments or information units. We would note which segment came first and then obtain the following data for each segment: the specific topic (several of which were precoded based on previous observations), themes (most were precoded), and sources used (mainly precoded, these refer to originators of information, claims, and impressions in the various reports).

Next come the visuals. Previous research and especially the articulation of the nature and significance of different media formats has enhanced understanding of the process of news selection and presentation of news reports (Altheide, 1985a, 1985b, 1985c; Altheide & Snow, 1991; Ericson, Baranek, & Chan, 1989). This category includes precoded distinctions as well as descriptions. In particular, it is important to distinguish the visuals by type: for example, videotape, file photos, graphics, actuals, and live coverage. Also, for each visual, it is necessary to know who is shown and where (scene), what action is shown (running, screaming, laughing), as well as other information about the visual, including what is different about it compared to other visuals noted. The protocol also includes segments for the researchers to make their own observations and comparisons, including cautionary remarks, such as about what one is uncertain.

The idea, then, is to "surround" the news report but to do it from "within" the report by examining the constituent parts. Akin to a paragraph, the news report has a unity, but its segments, akin to sentences, give it body. The reporters produce the segments, which are often edited together by others to produce the report. News organizations have established routines to accomplish producing the news, but the networks' approach changes if visuals, and particularly the all-important videotape, are not available. Recall that a central feature of the TV news format—especially the regularly scheduled evening news report—is that time and importance are marked with videotape, especially dramatic action tape. Thus, there is the interest in combat footage or Super Bowl highlights. From this perspective, what the journalist needs is exciting videotape for the evening newscast; their work entails access to the tape, the visuals. That defines air time, importance, and, to a remarkable degree, good work. If the access and the tape are not forthcoming, as was the case in Grenada, then we have a peculiar situation. On the one hand, some coverage must follow in that, after all, one's country is involved and viewers are interested in this sort of thing—in part because of the constant hype of conflict and violence that pervades news and entertainment programming. On the other hand, the other news organizations are likely to cover the war. There is the archival file and graphic materials to use, as was done with Grenada and Panama in particular.

The approach to document analysis relies on the researcher's protocol and categories and the capacity of comparisons to produce several differences in codes within each category. These codes can be expanded, reduced, or dropped altogether during the study as the investigator learns more about the data. This means that coding is influenced by a conceptual issue (e.g., accounts discussed in Chapter 4), but also that it is grounded to other similar data. Consistency of relevance and meaning emerges across the coding exercise. This is why a researcher becomes better and more informed during the research experience, and this suggests that a researcher acquires a kind of "reflexive relevance" and sensitivity to subtle distinctions in doing this work. So one can receive an orientation to data collection, but it is through interaction with other researchers as well as the research materials that one becomes a good data collector and coder. In a sense, you become a better researcher by doing the research.

By the same token, however, this does not mean that two or more data collectors cannot share perspectives and criteria. To the contrary, the training and orienting of people to do qualitative data analysis just requires more time and more involvement in the entire process of the research, from discussion of the conceptual rationale, to the exploration and selection of materials (documents), to reflection and discussion, to the establishment and refinement of protocols and categories—and their rationale and the coding process. The sense of the whole provides a context of understanding that will promote the best reflexive coding.

6. TRACKING DISCOURSE

The capacity to define the situation for self and others is a key dimension of social power. One reason to study mass media documents is to understand the nature and process by which a key defining aspect of our effective environment operates and to attempt to gauge the consequences. The media are consequential in social life. Numerous studies strongly suggest that public perceptions of problems and issues (the texts they construct from experience) incorporate definitions, scenarios, and language from news reports (Altheide & Snow, 1991; Bennett, 1988; Comstock, 1980; DeFleur & Ball-Rokeach, 1982, p. 244ff; Snow, 1983). What we call things, the themes and discourse we employ, and how we frame and allude to experience is crucial for what we take for granted and assume to be true. Simultaneously, we experience, reflect on that experience, and direct future experience. When language changes and new or revised frameworks of meaning become part of the public domain and are routinely used, then social life has been changed, even in a small way. This is why the topic of *discourse*—or the kinds of framing, inclusion, and exclusion of certain points of view—are important.

There are numerous approaches to the study of discourse. Indeed, several journals, crossing various disciplines—from the humanities to the social sciences—are devoted to discourse analysis (Grimshaw & Burke, 1994; Perinbanayagam, 1991; van Dijk, 1988; Weiler & Pearce, 1992; Wuthnow, 1992; Zhondang & Kosicki, 1993). Although there are many differences in some of the approaches, all share an assumption that symbolic representations are enmeshed in a context of other assumptions. My approach draws on many of these assumptions, but blends interpretive, ethnographic, and ethnomethodological approaches with media logic, particularly studies of news organizational culture, information technology, and communication formats.

New information technology, when combined with a theoretical and methodological approach to qualitative document analysis, can extend our capacity to study and understand public discourse. Previous work on news formats and "news codes" (Altheide, 1985b, p. 102ff) directs our attention away from the "intention" of the journalist who speaks about an event to questions about "what the news message looks like" or "what words and powerful cultural symbols are used in discussing the event."

I am particularly interested in how certain terminology and points of view arise and change or "travel" across the social environment. Studies of the news media by numerous scholars in several countries (see Appendix) suggest that a major news frame within which numerous reports are crafted is "fear." Fear is an important social problem because it often leads people to look for fear-reducing solutions, usually involving the state's use of force. Despite clear evidence showing that Americans today have a comparative advantage in terms of diseases, accidents, nutrition, medical care, and life expectancy, they perceive themselves to be at great risk and express clear fears about this. According to numerous public opinion polls, American society is a very fearful society, some believe "the most anxious, frightened society in history." Indeed, 78% of Americans think they are subjected to more risk today than their parents were 20 years ago, and a large source of this perception is crime news coverage, as illustrated in the following (Shaw, 1994):

> Why did many Americans suddenly decide last fall, for the first time, to tell national pollsters that crime is "the most important problem facing the country"? Could it have been because last year, for the first time, ABC, CBS and NBC nightly news programs devoted more time to crime than to any other topic?
>
> Several media critics think so; as a *Los Angeles Times* Poll showed early this year, people say their "feelings about crime" are based 65% on what they read and see in the media and 21% on experience. (p. 1)

These considerations, along with research on the mass media, raise the question of whether our public accounts of order and disorder—news reports (Ericson et al., 1989)—have undergone a qualitative change regarding the nature, use, and context of *fear*. For a media researcher, several questions that come to mind include, "What is the frequency and nature of 'fear' as it is presented in the news media, with what is fear associated, and has this changed?"

One way to approach these questions is through a project I call "tracking discourse," or following certain issues, words, themes, and frames over a period of time, across different issues, and across different news media. Tracking discourse is a qualitative document analysis technique that applies an ethnographic approach of content analysis to news information bases that are accessible through computer technology. Consistent with the approach discussed in earlier chapters, tracking discourse approaches meaning interactively and inductively with the materials. Tracking discourse helps expand qualitative analysis to an entire population of documents; with a foundation in depth and process, this approach can add additional avenues for exploration as well as breadth. Protocols, categories, and codes are informed by conceptual and theoretical issues, but their particular application emerges through interaction with the technology and the materials. The unit of analysis (e.g., the news article) can be modified with innovative sampling strategies (usually compatible with theoretical sampling) and even extended to an entire publication (e.g., the *Los Angeles Times* from 1985 to 1994).

The capacity to examine numerous documents with specific conceptually informed search terms and logic provides a new way of exploring documents, applying "natural experimental" research designs to the materials, as well as retrieving and analyzing individual documents qualitatively. Moreover, because the technology permits immediate access to an enormous amount of material, comparative exploration, conceptual refinement, data collection, and analysis can cover a longer time period than other technologies afforded.

NEXIS's and other information bases' search capacity provides a way to check on this because you can search for particular words and also bracket your searches by date—simply restricting your request to items that occurred before a certain date and after a certain date. The period between these dates, of course, is what you want. So to conduct a natural experiment to see if the use of *fear* has changed over time, all that was necessary was to select a library (which was "news"), a file (which could be the *Los Angeles Times* [*LAT*]), and a search request. In that I wanted to compare the use of fear at different times, I selected the earliest year in which the *LAT* was on line in NEXIS—1985—and compared it with the latest complete year—1994. The following search request was used (except with different dates): FEAR AND DATE (AFT 12/84 AND BEF 1/86). This was also done for 1994 (AFT 12/93 AND BEF 1/95) and the results are presented in Column A of the following.

In addition, a theoretical rationale about how issues travel across media formats suggested that the headlines may have changed over time to include more stories about fear. This was checked for each of the years using the same search, except each one was preceded with the "segment" headline: for example, HEADLINE (FEAR) AND DATE (AFT 12/84 AND BEF 1/86). Those results appear in Column B of the following.

A Articles in *LAT*	B Headlines in *LAT*
1985—4,579	1985—271
1994—7,440	1994—707
+62%	+161%

Two things stand out. First, in the *Los Angeles Times* at least, there was a notable increase—some 62%—in the use of the word *fear.* Second, there was an even larger increase in use of the word *fear* in headlines—161%.

These materials suggest that a qualitative shift in news reporting of fear occurred, particularly with headlines. But more information was needed to see how "thick" and widespread the shift may have been. To check on this, similar searches were conducted for other newspapers that had been on NEXIS for several years (this is a drawback for long looks back in time) as well as *ABC News* transcripts.

Further analysis of the materials indicated that the word *fear* is being used more often in public information (e.g., news reports, both print and TV). *Fear* is used as a verb, noun, and adverb. In many cases, it could be used synonymously with *concern*. Most important is the disproportionate way in which *fear* has moved into the headlines of newspapers and the leads of network news, at least at ABC. In each of the major newspapers examined, there was a larger increase in the word *fear* in headlines than in regular reports.

One other point should be noted: Obtaining the data for this analysis took less than 2 hours! Imagine, in less 2 hours a tentative answer has been provided to a research question potentially involving thousands of documents: "Has the word *fear* been used more often in news media reports over the last several years?" The answer is yes! But we still did not have a systematic way of appreciating what fear means in relation to the mass media. One way to approach this question using the LEXIS/NEXIS information base is to examine which problems and issues are more closely associated with fear over a period of time.

With the help of several graduate students (Dion Dennis, Morgan Doran, and Jen Ferguson), additional data were collected prior to natural experiment and an overview of those materials that were obtained using a similar search procedure helped clarify the research task for the experiment. A protocol was constructed for newspaper reports as well as TV news. Included in the categories of this protocol are date, year, page/section (or with TV news, story number in the newscast), location of *fear* or other synonyms (e.g., *afraid* or *threatened*), length of article (if a newspaper), length of report (time if a TV news report), the subject matter, and who or what is feared. In addition, another category, "miscellaneous," was added, which includes headline, key phrases, specific sources, most relevant sentences that help identify the report and give it its qualitative signature. As before, the materials were read or "filtered" through the protocol for appropriate coding and descriptive phrases.

Specifically, one thing we were interested in was how closely and how frequently *fear* was associated with other words, such as *crime* or *violence*, in headlines. The rationale for this approach is that the meaning of two words is suggested by their proximity and their association. Consider the example of *violence* and *crime* in the following three sentences:

1. An act of *violence* that might be regarded as a *crime* occurred Saturday night.
2. A *violent crime* occurred Saturday night.
3. A [*crime* (violence)] occurred Saturday night.

The first sentence treats both separately as nouns, but perhaps related. In the second sentence, violence is an adjective for crime, part of its description and meaning. But the third sentence shows what happens when terms are continually used together, often merging. This sentence suggests that crime has incorporated violence into its meaning, and the word *violent* might not even be used. Other studies have demonstrated how this happens with numerous social problems and issues. Not only is the event distorted by this coupling, but our capacity to deal with it in different ways also may be compromised. A similar coupling occurs when TV reports about crime and violence show individuals of certain racial and ethnic groups. Thus, TV visual formats can contribute to social definitions. Conversely, coupling may not occur between words and topics if they have traditionally been viewed as quite separate. There is reason to suspect that this is part of the difficulty in convincing people that domestic violence is "real violence" and also a "crime." The notions of "family" and "crime" and "violence" have seldom appeared within close proximity in routine news reports until fairly recently.

Only a small section of the study is reported here. Analysis of the headlines helps clarify the different meanings and associations of the word *fear* as presented in public discourse. The headlines for 1992 in the *Los Angeles Times* suggested that fear is volatile. Among the points to draw from research to date include the following: (a) The discourse of fear has shifted over a several-year period to include certain problems and issues; (b) the closest association with fear in 1985 was AIDS, and this shifted to violence in 1992; and (c) certain problems and issues are more closely associated with fear than others. What this suggests is that in 1992, the *Los Angeles Times* headlines presented different associations for and meanings of fear, and the problems or issues most closely affiliated with fear were violence, crime, and drugs. An archetype headline combines all three in the following *LAT* headline (1992, July 5):

Crime Cut by Bikes, Barricades, Law Enforcement: The Neighborhood was a *Drug* Haven Until New Tactics Were Implemented. The Two-Wheel Patrols and Barriers Have Made a Dent in the Danger But Some Residents Still Live in *Fear*

The increasingly common association of several of these terms with another menace, "gang," is illustrated in the following headline: "Anaheim *Fears Drug* Turf Wars Among *Gangs*" (1992, February 6).

Careful reading and analysis (along the lines suggested in Chapter 3) of a sample of articles selected in this kind of search helps reveal the broad patterns of increased prominence of fear as well as topics it is associated with and the sources helping journalists make these assertions. Additional qualitative analysis of the articles indicates that certain sources (e.g., law enforcement and other formal agents of control) are more likely to be associated with reports about crime, violence, and drugs that are associated with fear (e.g., "Task Force to Offer Plan by June: Compton Targets Drugs, Killings," *LAT*, 1986, February 2). On the other hand, qualitative analysis of reports involving cancer and the environment suggests a difference in focus and source. The focus is often on new information, research, and controversies suggesting ambiguity of definition and the very nature of the problem: for example, "Officials Try to Calm Cancer Fears; Data Shows Disease Rate Average Near Toxic Waste Dump" (*LAT*, 1989, August 4) and "State Seeks Health Files of Rockwell Employees; Investigation: Officials Say They Will Gather Records For a Study to Determine Whether Workers Have An Unusual Rate of Bladder Cancer or Other Disease" (*LAT*, 1991, February 9). Agencies and experts other than those affiliated with law enforcement and crime are likely to be the source of information (e.g., scientists, medical associations).

Notwithstanding debates that will surround subsequent interpretations of these and other materials derived from tracking discourse, it is apparent that information bases, when approached with a broad qualitative perspective, can provide rich insights, research questions, and innovative answers. This processual approach to systematic analysis of documents suggested that the term *fear* can be tracked in document bases to illuminate specific referents and meanings of *fear*, including relative frequencies, association with *fear*, as well as the most common news sources (organizations and agencies) that are affiliated with and "own" certain meanings of *fear*. The upshot is that qualitative document analysis has been given a rebirth by information technology. The challenge is for innovative researchers to boldly experiment and develop methodological applications worthy of these new developments.

7. FIELD NOTES AND OTHER DATA

Virtually all research involves documents and most of these can be analyzed qualitatively. Although the focus in this book has been on *primary documents,* this chapter examines how to use many of these principles to work with *secondary documents,* including interviews, field notes from observations, and records. Content analysis, then, is basic to most research, regardless whether the original data were derived from other documents such as newspapers, TV news reports, personal observations, or interviews. The researcher develops a record of some kind; if this is retrievable and subject to analysis, it is a document of research—a kind of "account" of some interaction or activity relevant to one's study. In the case of qualitative research, these documents are often written text or narrative—for example, accounts of what took place, descriptions, summaries of observations, or interviews. The only question (actually there are two) that remains then, is how are these documents (notes) to be organized and analyzed and when and how should one go about this? The brief comments to follow are intended to provide some answers to these queries. Although there are different approaches to these questions (Berg, 1989, pp. 105-127), the following guidelines are based on my own work and numerous students' projects. They should be regarded as some "minimalist" considerations in planning and conducting a project involving field notes, but in most cases a researcher will want to supplement these with more detailed analytical principles of various theoretical approaches—for example, grounded theory (Berg, 1989; Strauss, 1987).

A basic consideration in organizing one's field notes is that the data that are originally recorded determine the quality of whatever is later coded. Coding cannot improve substantially the quality of one's data, but it can help get the most out of it for a report. Ideally, very early in the research process an investigator would have an awareness of the interaction between substantive interests, data collection, organization, and coding. A few comments about ethnography and the research process can help to clarify what kinds of things should be noted and recorded in one's notes.

The process of conducting field observations and interviews is quite complex, and there are a number of fine references (Denzin, 1989; Douglas, 1976; Johnson, 1975; Jorgensen, 1989). The process I wish to stress here is for the researcher to be as close as possible to the setting and its activities under investigation. Learn the language, perspectives, routines, and practical considerations to determine "how" people do things and "what" they actually do. Experience them and avoid, until much later in the study, the question of "why" people do the things they do. This will usually become clear later. Meanwhile, one takes notes, and the best notes are descriptions and very rich in detail. Notes should not be taken to be significant or important for others, but simply as a good chronicle of what was done, seen, heard, and even felt.

Notes should be completed as quickly as possible after leaving the setting or when the opportunity permits. In general, I have found that it is a good idea to avoid taking notes in the presence of the members one is studying. It is better to jot down on a note card an occasional "key word" or phrase that will jog one's memory later. It is more important to not miss the various social activities than to chronicle some in favor of others. Notes can also include questions to oneself, points of confusion. Never avoid putting something down because it is not well enough understood, even though many of your initial observations will later be revised and even rejected as you understand more about the setting and activities.

Note cards can be consulted in a timely fashion to complete and round out one's field notes for the day. These should include details on relevant items, including a list of socially relevant categories (as we anticipate coding!) listed below. Most researchers will develop their own style of note taking and recording, but many agree that recording them in a chronological manner can be helpful. My students agree that it is also helpful to make comments in a separate part of one's notes for the day about the research process, difficulties, uncertainties, and practical problems—for example, "I couldn't pay attention to what X was saying because I had to go to the bathroom."

Accounting for Substance

Even though field notes are very specific and should not be enslaved to abstract concepts and theoretical issues in the day-to-day observations, there are some guidelines of the kinds of things that should be included. The following are generic topics that should be included in ethnographic reports. Each of these can be regarded as both a guideline for data collection and the actual development of research documents—for example, field notes, as well as categories and codes for subsequent data analysis.

The contexts, history, physical setting, and environment
Number of participants, key individuals
Activities
Schedules, temporal order

Division of labor, hierarchies
Routines and variations
Significant events, origins and consequences
Members' perspectives and meanings
Social rules and basic patterns of order

Different researchers will develop individual approaches, of course, but the idea is that materials relevant to these dimensions would be collected and recorded in the course of an ethnographic study. This does not mean, however, that a researcher should begin and be guided by such abstract concepts. Rather, the best ethnographic research is always very specific and descriptive. The relevance of what one should describe and about what information one should obtain will become more clear as the researcher consults previous studies, but most important, becomes immersed in the setting, and engages in the members' world and activities. The point is that after several periods of observation and data collection, an investigator can simply review notes with these categories as a rough checklist of some relevant considerations. In this sense, the above categories and others that most researchers will discover become potential codes that can be inserted into one's notes as the study progresses. Such "tags" in the notes also provide a way for the investigator to check the relative amount of attention and materials available on different dimensions of the research project.

What follows is an example of preliminary codes or tags I have attached to notes provided by Berg (1989, p. 75). I use Berg's example rather than my own to illustrate the versatility of the coding categories listed previously; practically anyone's data can be organized by using them. The codes or tags used are marked in bold type with an asterisk, ***bold.**

[TIME: 9:40 I left the meeting with the parents' advisory group and Barry a few minutes past 9:00 p.m. I went directly to Eddie's Bar. After parking my car directly in front of the bar, I started toward the door. I immediately noticed Olaf hurrying in. I followed him inside the bar. ***setting** The inside of Eddie's consists of three separate rooms. The first room one enters is the main bar room. It is set up like a traditional neighborhood bar: one long bar counter (to the right of the entrance), a television up on the wall at the far end of the bar, and a few booths set along the left side of the room. To the immediate left, as one faces the wall with the booths, there is a doorway leading to a small room (approximately ten by ten). This room contains a small billiard table, a Foosball table, and five or six chairs and small tables. There are two large stereo speakers on the wall to the extreme left and a small D.J. booth in the far left corner. The lights were very dim in this room, and the music being played (rock and roll) was very loud. ***atmosphere**

*participants When I first walked into the main bar room, I noticed four very young looking kids seated in the first booth (they may have been fourteen or fifteen years old). I didn't know any of them. I continued looking for faces I recognized. At the bar counter several older men stood or sat on stools, drinking. Toward the back of the main bar room, where there were several small tables, I could see several more young-looking kids sitting around, some on the tables themselves, and others in chairs. I still saw no one I knew.

I walked through the entrance into the large room off the rear of the bar room. Sitting on what appeared to be a bar counter (much smaller than the one in the main bar room and not in use this evening) were three girls I recognized from Oxford High School. I smiled at them, waved a greeting and said hello as I approached them. One of the girls (the one seated in the middle) leaned over to the girl to her left and audibly whispered, "Do you know this guy?" The girl being asked nodded her head yes, and said, "Yeah, I met him at the school play rehearsal the other day." I walked on past these three girls as I spotted Audrey Miller drinking a beer and sitting on top of one of the other small tables. Audrey was sitting with her right arm draped over the shoulder of some guy sitting next to her (I didn't know him). As I moved closer to her, she looked up and said, "Hello, Bruce." Her eyes widened, and she appeared a little surprised to see me. She got up off the table and walked over to me. ***view of researcher** She asked if I was there doing research or just out socializing. ***member perspective** I told her I was doing research. She remarked, "Well, you've certainly come to the right place, this whole room is filled with Oxford kids." ***identity** She was just slightly slurring her words, suggesting that the beer she held in her right hand was not her first. She said, "I'll see you later," and walked back to the group of kids with whom she had been sitting.] (From Berg, 1989. *Qualitative Research Methods for the Social Sciences.* Copyright © 1989 by Allyn and Boacon. Reprinted by permission.)

The codes I have inserted in these notes are not the only possible ones, of course, because one's research focus will influence this. Regardless, the important point is that this brief description of a research setting has several clear components of an ethnographic study, and these can be pulled out and combined with similar codes to help analyze and then write sections of a report.

Good descriptive data on various dimensions of the activity and setting under investigation make coding much easier. Research documents should contain materials about the substantive interest as well as the research process. As noted above, planning for this will facilitate the later coding and organization phases of the research. All researchers should provide an account of the research act. This is important for reasons of credibility as well as validity (although some investigators are no longer concerned about this). It is also relevant for substantive theory as research methodology (Glaser & Strauss, 1967).

Another useful strategy in coding field materials is to develop gerunds from the research materials. Gerunds are "verbal nouns" usually identified by ending in *ing*. They can be very descriptive while also referring to an activity the researcher deems relevant. For example, in my studies of television activities, numerous descriptions in my notes were tagged with gerunds such as, *editing, filming, complaining about story assignments, goofing around, getting screwed* (usually by the news director). When the more action-oriented but still very generic *doing* is considered, then the field researcher is on track of the symbolic interactionist perspective, which essentially views social activities as accomplishments with a process. Thus, "doing nothing" is a meaningful category if described in a research setting, as is "hanging out." Everyday routines in most settings can be captured in this way. Even if they are not originally entered as data, however, these are viable coding categories for most studies. These dimensions provide a quasi-template for an investigator and a prospective reader of the report to understand what contributes to the definition of the situation, its nature, character, origin, and consequences.

Accounting for Ourselves

A key part of the ethnographic ethic is how we account for ourselves. Good ethnographies show the hand of the ethnographer. The effort may not always be successful, but there should be clear tracks that the attempt has been made. Experience suggests that there is a minimal set of problem areas that are likely to be encountered in most studies. The following does not offer a solution to the problems that will follow but only suggest that these can provide a focus for providing a broader and more complete account of the reflexive process through which something is understood (Altheide, 1976; Denzin, 1989; Douglas, 1976; Johnson, 1975). Such information enables the reader to engage the study in an interactive process that includes seeking more information, contextualizing findings, reliving the report as the playing out of the interactions between the researcher, the subjects, and the topic in question.

As with the substantive categories (potential codes), the research process guidelines can be treated similarly, inserted as tags into relevant sections of notes. Suggested items for locating and informing the role of the researcher vis-à-vis the phenomenon include a statement about topics previously delineated in other work (Altheide, 1976, p. 197ff):

Entree—organizational and individual
Approach and self-presentation
Trust and rapport
Researcher's role and way of fitting in
Mistakes, misconceptions, surprises
Types and varieties of data
Data collection and recording
Data coding and organization
Data demonstration and analytical use

Because these dimensions of ethnographic research are so pervasive and important for obtaining truthful accounts, they should be implicitly or explicitly addressed in the report. Drawing on such criteria enables the ethnographic reader to approach the ethnography interactively and critically to ask, What was done? How was it done? What are the likely and foreseen consequences of the particular research issue? In what way was it was handled by the researcher? No study avoids all of these problems, although few researchers give a reflexive account of their research problems and experience. There are other potential problems that should be considered in one's final report and, therefore, chronicled in field notes. Douglas (1976) cautions us about problems of communication with informants: misinformation, evasions, lies, fronts, taken-for-granted meanings, problematic meanings, and self-deceptions. For example, the field notes from Berg (above) could be coded throughout the study as the researcher becomes more aware of things missed or misunderstood. Attending to these issues in a study does not make the study more truthful but only means that the truth claims of the researcher can be more systematically assessed by readers who share a concern with the relationship between *what* was observed and *how* it was accomplished.

Finally, these suggestions are not intended to describe grounded theory or any other abstract view of data gathering and coding, although they may be applicable, depending on the project and the researcher. It has been my experience that the analytic inductive approach to grounded theory can be useful in general—but the detailed coding guidelines make the data organizing process more difficult than it need be—and that some of the coding rules lead researchers to essentially ignore certain data that do not fit the pursuit of an emerging category. My inclination is to include all the data in one's analysis, although some may receive more attention than others. Thus, I prefer the term *emergent coding* for what has been described. The first goal is to give an accurate account of the complexities underlying the simplest scenario, and only then is more abstract theoretical sense to be derived, including relating this study to others. By using certain indexing technologies such as Zyindex (referred to in a previous chapter), your "documents of documents," your theoretical accounts of mundane but very important everyday living will continue to emerge, be reborn, and lead to new insights.

Over the years, I have attempted to integrate the range of findings from studies on the mass media, and especially TV news, with the nature and significance of media in general in social life. I have examined non-mass-media institutions, settings, and practices to catch conceptually informed glimpses of the impact of certain mass-media effects on a wide range of activities, as well as to further clarify the relevance of a more general focus on the range of media that influence temporal and spatial features of what appear at first glance to be nonmediated occasions. Such an approach yielded "format" as a common concept to a number of fine studies of mass media as well as other types of mediation. Looking for instances of mediation in situations that may not conventionally be associated with media principles and theory has led to settings and issues involving social definitions and applications of "justice," including TV coverage of court-room activity, the use of "keyboards" and other terminals by police officers and other criminal justice agents (Altheide, 1985b), and a study of TV viewers' requests for assistance from an action line "troubleshooter." These experiences and capacity to reexamine my data helped promote a more expansive conception of mediation that goes well beyond my initial focus on formats: An *ecology of communication* involves the nondeterministic influence of information technology and formats on social activities (Altheide, 1995). In brief, ethnography offers a perspective for analysis of human action in the field and in documents; the key is to reconceptualize the latter as the former. Documents remain to be discovered through the research process, a process that will undoubtedly encourage other researchers to reflect on the experience and materials and offer yet other ways for studying documents of our lives.

82

APPENDIX

Selected Studies of News Organizations

Adams, W. (Ed.). (1982). *Television coverage of international affairs.* Norwood, NJ: Ablex.

Altheide, D. L. (1976). *Creating reality: How TV news distorts events.* Beverly Hills, CA: Sage.

Batscha, R. M. (1975). *Foreign affairs news and the broadcast journalist.* New York: Praeger.

Bennett, L. (1988). *News: The politics of illusion* (2nd ed.). New York: Longman.

Braestrup, P. (1978). *Big story: How the American press and television reported and interpreted the crisis of Tet in 1968 in Vietnam and Washington.* Garden City, NY: Anchor.

Chibnall, S. (1977). *Law-and-order news.* London: Tavistock.

Epstein, E. J. (1973). *News from nowhere.* New York: Random House.

Ericson, R. V., Baranek, P. M., & Chan, J. B. L. (1987). *Visualizing deviance: A study of news organization.* Toronto: University of Toronto Press.

Ericson, R. V., Baranek, P. M., & Chan, J. B. L. (1989). *Negotiating control: A study of news sources.* Toronto: University of Toronto Press.

Ericson, R. V., Baranek, P. M., & Chan, J. B. L. (1991). *Representing order: Crime, law, and justice in the news media.* Toronto: University of Toronto Press.

Fishman, M. (1980). *Manufacturing the news.* Austin: University of Texas Press.

Gans, H. J. (1979). *Deciding what's news.* New York: Pantheon.

Gitlin, T. (1980). *The whole world is watching.* Berkeley, CA: University of California Press.

Glasgow University Media Group. (1976). *Bad news.* London: Routledge and Kegan Paul.

Schlesinger, P. (1978). *Putting "reality" together.* London: Constable.

Schlesinger, P., Murdock, P. G., & Elliott, P. (1983). *Televising "terrorism": Political violence in popular culture.* London: Comedia Publishing Group.

Schudson, M. (1978). *Discovering the news: A social history of American newspapers.* New York: Basic Books.

Surette, R. (1992). *Media, crime and criminal justice: Images and realities.* Pacific Grove, CA: Brooks/Cole.

Tuchman, G. (1978). *Making news: A study in the construction of reality.* New York: Free Press.

Selected Studies of Entertainment Television

Cantor, M. G. (1980). *Prime-time television: Content and control.* Beverly Hills, CA: Sage.

Combs, J. (1984). *Polpop: Politics and popular culture in America.* Bowling Green, OH: Bowling Green University Press.

Comstock, G. (1980). *Television in America.* Beverly Hills, CA: Sage.

Gitlin, T. (1983). *Inside prime time.* New York: Pantheon.

Sklar, R. (1980). *Prime-time America.* New York: Oxford University Press.

Snow, R. P. (1983). *Creating media culture.* Newbury Park, CA: Sage.

Selected Studies in Social Science Techniques and Methods

Altheide, D. L., & Johnson, J. M. (1993). Tacit knowledge: The boundaries of experience. In N. K. Denzin (Ed.), *Studies in symbolic interaction* (Vol. 13, pp. 53-57). Greenwich, CT: JAI.

Altheide, D. L., & Johnson, J. M. (1994). Criteria for assessing interpretive validity in qualitative research. In N. K. Denzin & Y. S. Lincoln (Eds.), *Handbook of qualitative research* (pp. 485-499). Newbury Park, CA: Sage.

Brown, R. H. (Ed.). (1992). *Writing the social text: Poetics and politics in social science discourse.* New York: Aldine.

Harper, D. (1987). *Working knowledge: Skill and community in a small shop.* Berkeley: University of California Press.

Hill, N. R. (1993). *Archival strategies and techniques.* Thousand Oaks, CA: Sage.

Lindlof, T. R. (1995). *Qualitative communication research methods.* Thousand Oaks, CA: Sage.

Manning, P. (1987). *Semiotics and fieldwork.* Newbury Park, CA: Sage.

Manning, P. (1988). *Symbolic communication: Signifying calls and policework.* Cambridge: MIT Press.

Miles, M. B., & Huberman, A. M. (1995). *Qualitative data analysis* (2nd ed.). Thousand Oaks, CA: Sage.

Webb, E. J., Campbell, D. T., Schwartz, R. D., & Sechrest, L. (1966). *Unobtrusive measures: Nonreactive research in the social sciences.* Chicago: Rand McNally.

REFERENCES

Adams, W., & Schreibman, F. (Eds.). (1978). *Television network news: Issues in content research.* Washington, DC: George Washington University.

Altheide, D. L. (1976). *Creating reality: How TV news distorts events.* Beverly Hills, CA: Sage.

Altheide, D. L. (1981). Iran vs. U.S. TV news: The hostage story out of context. In W. Adams (Ed.), *TV coverage of the Middle East* (pp. 128-158). Norwood, NJ: Ablex.

Altheide, D. L. (1982). Three-in-one news: Network coverage of Iran. *Journalism Quarterly, 48,* 476-490.

Altheide, D. L. (1985a). Format and ideology in TV news coverage of Iran. *Journalism Quarterly, 62,* 346-351.

Altheide, D. L. (1985b, Fall). Keyboarding as a social form. *Computers and the Social Sciences, 1,* 97-106.

Altheide, D. L. (1985c). *Media power.* Beverly Hills, CA: Sage.

Altheide, D. L. (1987). Ethnographic content analysis. *Qualitative Sociology, 10,* 65-77.

Altheide, D. L. (1992). Gonzo justice. *Symbolic Interaction, 15,* 69-86.

Altheide, D. L. (1995). *An ecology of communication: Cultural formats of control.* Hawthorne, NY: Aldine de Gruyter.

Altheide, D. L., & Snow, R. P. (1979). *Media logic.* Beverly Hills, CA: Sage.

Altheide, D. L., & Snow, R. P. (1991). *Media worlds in the postjournalism era.* Hawthorne, NY: Aldine de Gruyter.

Anaheim fears drug turf wars among gangs. (1992, February 6). *Los Angeles Times,* p. A1.

Ball, M. S., & Smith, G. W. H. (1992). *Analyzing visual data.* Newbury Park, CA: Sage.

Bennett, W. L. (1988). *News: The politics of illusion* (2nd ed.). New York: Longman.

Bennett, W. L., & Paletz, D. L. (Eds.). (1994). *Taken by storm: The media, public opinion, and U.S. foreign policy in the Gulf War.* Chicago: University of Chicago Press.

Berelson, B. (1966). Content analysis in communication research. In B. Berelson & M. Janowitz (Eds.), *Reader in public opinion and communication* (pp. 260-266). New York: Free Press.

Berg, B. L. (1989). *Qualitative research methods for the social sciences.* Boston: Allyn & Bacon.

Berger, A. A. (1981). Semiotics and TV. In R. P. Adler (Ed.), *Understanding television* (pp. 91-114). New York: Praeger.

Berger, A. A. (1982). *Media analysis techniques.* Newbury Park, CA: Sage.

Berger, P., & Luckmann, T. (1967). *The social construction of reality.* New York: Anchor.

Brissett, D., & Edgley, C. (Eds.). (1990). *Life as theater* (2nd ed.). New York: Aldine de Gruyter.

Burke, K. (1962). *A grammar of motives and a rhetoric of motives.* New York: World Publishing.

Chayako, M. (1993). What is real in the age of virtual reality? "Reframing" frame analysis for a technological world. *Symbolic Interaction, 16,* 171-182.

Cicourel, A. (1964). *Method and measurement in sociology.* New York: Free Press.

Comstock, G. (1980). *Television in America.* Beverly Hills, CA: Sage.

Combs, J. (1984). *Polpop: Politics and popular culture in America.* Bowling Green, OH: Bowling Green University Press.

Crime cut by bikes, barricades, law enforcement: The neighborhood was a drug haven until new tactics were implemented. (1992, July 5). *Los Angeles Times,* p. 1.

DeFleur, M. L., & Ball-Rokeach, S. (1982). *Theories of mass communication* (4th ed.). New York: Longman.

Denzin, N. K. (1989). *The research act.* Chicago: Aldine.

Denzin, N. K., & Lincoln, Y. S. (1994). *Handbook of qualitative research.* Newbury Park, CA: Sage.

Douglas, J. D. (1976). *Investigative social research.* Beverly Hills, CA: Sage.

Epstein, E. J. (1973). *News from nowhere.* New York: Random House.

Ericson, R. V., Baranek, P. M., & Chan, J. B. L. (1989). *Negotiating control: A study of news sources.* Toronto: University of Toronto Press.

Feldman, M. (1994). *Strategies for interpreting qualitative data.* Thousand Oaks, CA: Sage.

Ferguson, J. (1995, April). *Newspaper accounts of police-officer involved shootings.* Paper presented at the Annual Meeting of the Western Social Science Association, Oakland, CA.

Fields, E. E. (1988). Qualitative content analysis of television news: Systematic techniques. *Qualitative Sociology, 11*(3), 183-193.

Fishman, M. (1980). *Manufacturing the news.* Austin: University of Texas Press.

Gitlin, T. (1980). *The whole world is watching.* Berkeley: University of California Press.

Glaser, B., & Strauss, A. (1967). *The discovery of grounded theory.* Chicago: Aldine.

Glasgow Media Group. (1985). *War and peace news.* Milton Keynes, UK: Open University Press.

Goffman, E. (1974). *Frame analysis.* New York: Harper and Row.

Graber, D. A. (Ed.). (1984). *Media power in politics* (pp. 239-250). Washington, DC: Congressional Quarterly Press.

Grimshaw, A. D., & Burke, P. J. (1994). *What's going on here: Complementary studies of professional talk.* Norwood, NJ: Ablex.

Hammersley, M., & Atkinson, P. (1983). *Ethnography: Principles in practice.* New York: Tavistock.

Henry, G. T. (1990). *Practical sampling.* Thousand Oaks, CA: Sage.

Hessler, R. M. (1992). *Social research methods.* Los Angeles: West.

Holsti, O. R. (1969). *Content analysis for the social sciences and humanities.* Reading, MA: Addison-Wesley.

Huberman, A. M., & Miles, M. B. (1994). Data management and analysis methods. In N. K. Denzin & Y. S. Lincoln (Eds.), *Handbook of qualitative research* (pp. 428-444). Newbury Park, CA: Sage.

Johnson, J. M. (1975). *Doing field research.* New York: Free Press.

Jorgensen, D. L. (1989). *Participant observation: A methodology for human studies.* Newbury Park, CA: Sage.

Judge orders lessons in Holocaust. (1990, May 9). *St. Petersburg Times,* p. 10A.

Kelle, U. (1995). *Computer-aided qualitative research.* Newbury Park, CA: Sage.

Krippendorf, K. (1978). The expression of values in political documents. *Journalism Quarterly, 55,* 510-518.

Lang, K., & Lang, G. E. (1968). *Politics and television.* Chicago: Quadrangle.

Manning, P. (1992). *Organizational communication.* New York: Aldine de Gruyter.

Manning, P., & Cullum-Swan, B. (1994). Narrative, content and semiotic analysis. In N. K. Denzin & Y. S. Lincoln (Eds.), *Handbook of qualitative research* (pp. 463-484). Newbury Park, CA: Sage.

McCormack, T. (Ed.). (1982). Content analysis: The social history of a method. *Studies in communications: Vol. 2. Culture, code and content analysis* (pp. 143-178). Greenwich, CT: JAI.

Mercer, D. (1987). The media on the battlefield. In D. Mercer, G. Mungham, & K. Williams (Eds.), *The fog of war* (pp. 1-16). London: Heinemann.

Mercer, D., Mungham, G., & Williams, K. (1987). *The fog of war*. London: Heinemann.

Molotch, H., & Boden, D. (1985). Talking social structure: Discourse, domination and the Watergate hearings. *American Sociological Review, 50*(3), 273-288.

Morrison, D. E. (1992). *Television and the Gulf War*. London: John Libbey.

Morrison, D. E., & Tumber, H. (1988). *Journalists at war: The dynamics of news reporting during the Falklands conflict*. Beverly Hills, CA: Sage.

Mungham, G. (1987). Grenada: News blackout in the Caribbean. In D. Mercer, G. Mungham, & K. Williams (Eds.), *The fog of war* (pp. 291-310). London: Heinemann.

NBC Special Report. (1980, January 16). Crisis in Iran: 1 year after the Shah, day 75.

Officials try to calm cancer fears: Data shows disease rate average near toxic waste dump. (1989, August 4). *Los Angeles Times*, pt. 2, p. 10.

Perinbanayagam, R. (1991). *Discursive acts*. Hawthorne, NY: Aldine de Gruyter.

Pfaffenberger, B. (1988). *Microcomputer applications in qualitative research*. Newbury Park, CA: Sage.

Plummer, K. (1983). *Documents of life: An introduction to the problems and literature of a humanistic method*. Sydney, Australia: Allen and Unwin.

Rau, R. A. (1993). *The role of sources in news coverage of the Don Harding execution*. Unpublished master's thesis, Arizona State University, Tempe, AZ.

Richards, T. J., & Richards, L. (1994). Using computers in qualitative research. In N. K. Denzin & Y. S. Lincoln (Eds.), *Handbook of qualitative research* (pp. 445-462). Newbury Park, CA: Sage.

Scales of justice: Teacher who molested 2 girls forfeits piano. (1994, May 20). *Los Angeles Times*, p. 11.

Schutz, A. (1967). *The phenomenology of the social world* (G. Walsh & F. Lenhert, Trans.). Evanston, IL: Northwestern University.

Schwartz, H., & Jacobs, J. (1979). *Qualitative Sociology*. New York: Free Press.

Scott, M. B., & Lyman, S. M. (1968). Accounts. *American Sociological Review, 33*, 46-62.

Shaw, D. (1994, September 11). Headlines and high anxiety: We're safer and healthier than ever—and also more afraid of what we eat, drink and breathe. *Los Angeles Times* (Home Edition), p. 1.

Snow, R. P. (1983). *Creating media culture*. Newbury Park, CA: Sage.

Starosta, W. J. (1984). Qualitative content analysis: A Burkean perspective. In W. B. Gudykunst & Y. Y. Kim (Eds.), *Methods for intercultural communication* (pp. 185-194). Beverly Hills, CA: Sage.

State seeks health files of Rockwell employees. (1991, February 9). *Los Angeles Times*, pt. v, p. 3.

Strauss, A. (1987). *Qualitative analysis for social scientists*. New York: Cambridge University Press.

Strauss, A., & Corbin, J. (1990). *Basics of qualitative research: Grounded theory procedures and techniques*. Newbury Park, CA: Sage.

Task force to offer plan by June: Compton targets drugs, killings. (1986, February 2). *Los Angeles Times*, pt. 9, p. 2.

Thief broadcasts it. (1989, June 18). *The Arizona Republic*.

87

Thorpe, J. (1984, January 5). Fat guy told to diet or go to big house. *Tempe Daily News.*
Tuchman, G. (1978). *Making news: A study in the construction of reality.* New York: Free Press.
van Dijk, T. A. (1988). *News as discourse.* Hillsdale, NJ: Lawrence Erlbaum.
Weiler, M., & Pearce, W. B. (1992). *Reagan and public discourse in America.* Tuscaloosa: University of Alabama Press.
Weitzman, E. A., & Miles, M. B. (1995). *Computer programs for qualitative data analysis.* Thousand Oaks, CA: Sage.
Wuthnow, R. (Ed.). (1992). *Vocabularies of public life: Empirical essays in symbolic structure.* New York: Routledge.
Zhondang, P., & Kosicki, G. (1993). Framing analysis: An approach to news discourse. *Political communication, 10,* 55-69.

ABOUT THE AUTHOR

DAVID L. ALTHEIDE is Regents' Professor in the School of Justice Studies at Arizona State University, where he has taught since receiving his PhD from the University of California, San Diego, in 1974. His research interests in mass media, social control, and qualitative research methods are recorded in several books, including award-winning works such as *Creating Reality: How TV News Distorts Events* (1976), 1986 winner of the Premio Diego Fabbri Award, and *Media Power* (1985), winner of the Charles Horton Cooley Award of the Society for the Study of Symbolic Interaction, along with *Bureaucratic Propaganda* (coauthored with John M. Johnson), and *Media Logic* and *Media Worlds in the Postjournalism Era* (coauthored with Robert P. Snow). As the recipient of Arizona State University's Graduate College Distinguished Researcher Award for 1990-1991, he studied the role of the mass media in peace and war and presented some initial findings from this project at an international conference sponsored by the Fulbright Commission at the University of Nottingham, England (September, 1992). He has just completed another book, *An Ecology of Communication: Cultural Formats of Control* (1995), while serving as President of the Society for the Study of Symbolic Interaction. Altheide is married (Carla) and has two children, Tasha, a graduate student at the University of Arizona, and Tod, who is attending San Diego State University.